THERMAL PROCESSING OF READY-TO-EAT MEAT PRODUCTS

THERMAL PROCESSING OF READY-TO-EAT MEAT PRODUCTS

C. Lynn Knipe

Robert E. Rust

WILEY-BLACKWELL

A John Wiley & Sons, Ltd., Publication

Edition first published 2010

Blackwell Publishing was acquired by John Wiley & Sons in February 2007. Blackwell's publishing program has been merged with Wiley's global Scientific, Technical, and Medical business to form Wiley-Blackwell.

Editorial Office
2121 State Avenue, Ames, Iowa 50014-8300, USA

For details of our global editorial offices, for customer services and for information about how to apply for permission to reuse the copyright material in this book, please see our website at www.wiley.com/wiley-blackwell.

Library of Congress Cataloging-in-Publication Data

Knipe, C. Lynn.
Thermal processing of ready-to-eat meat products / C. Lynn Knipe, Robert E. Rust.
p. cm.
Includes bibliographical references and index.
ISBN 978-0-8138-0148-3 (hardback : alk. paper) 1. Food–Microbiology. 2. Food–Effect of heat on. 3. Industrial microbiology–Safety measures. 4. Meat–Preservation. I. Rust, Robert E. II. Title.
QR117.K55 2010
664.001′579–dc22 2009015160

A catalog record for this book is available from the U.S. Library of Congress.

Set in 11/13 pt Times by Aptara® Inc., New Delhi, India
Printed in Singapore by Markono Print Media Pte Ltd

1 2010

Dedication

Erwin Waters had retired from marketing, installing and troubleshooting thermal processing units (i.e., smoke houses) and had established himself as a consultant to the meat industry. Of particular interest to Erwin were the cooking and cooling processes and the emerging regulatory requirements related to preventing survival and growth of *Listeria monocytogenes* and *Clostridium perfringens* in and on ready-to-eat (RTE) meat products. Although it was job security for him, Erwin, always ready to share his expertise, saw an industry need for good technical information related to the thermal processing of RTE meat products. There were numerous meat processing courses available to the industry but none that specifically addressed thermal processing.

In the fall of 1999, Erwin contacted Bob Rust and Lynn Knipe to ask if they were interested in helping to develop such a course. As a result of that initial contact, the first Thermal Processing of Ready-to-Eat Meat Products short course was planned for spring of 2000 at Ohio State University. This course was designed to bring in experts in the areas of food microbiology, food engineering, regulatory requirements, sanitation, food science, and various heat transfer methods to orient and update meat industry employees.

In spite of his efforts in getting this course started, Erwin was only present for the first course in late February of 2000. His health had declined such that he had to cancel his appearance at the second course in March 2001. Erwin Waters passed away in December 2001.

There was never any question about the continuation of the course that Erwin had initially proposed. The surviving organizers of this course have attempted to improve and expand this course, following the spirit of Erwin's original proposal. Hopefully, Erwin would be pleased with how this has developed over the past ten years.

A couple of years ago, Wiley-Blackwell approached Lynn Knipe and Bob Rust about the possibility of publishing a reference book on the content of this course. This first edition is the result of Erwin Water's initiative and ten years of accumulated expertise related to the thermal processing of RTE meat products.

Contents

Contributors

Donald Burge
Director of R&D
Gold'nPlump Poultry
St. Cloud, MN

Mary Kay Folk
The Ohio State University
Food Science and Technology
Columbus, OH

Darrell Horn
Blentech Corporation
Santa Rosa, CA

Andy Hwang
U.S. Department of Agriculture
Agricultural Research Service
Eastern Regional Research Center
Microbial Food Safety Research Unit
600 East Mermaid Lane
Wyndmoor, PA

Vijay Juneja
U.S. Department of Agriculture
Agricultural Research Service
Eastern Regional Research Center
Microbial Food Safety Research Unit
600 East Mermaid Lane
Wyndmoor, PA

C. Lynn Knipe
The Ohio State University
Food Science and Technology
Columbus, OH

David Kramer
Sara Lee Corporation
Cincinnati, OH

Bradley P. Marks
Michigan State University
East Lansing, MI

Aubrey F. Mendonca
Department of Food Science and Human Nutrition
Iowa State University
Ames, IA

Robert E. Rust
Iowa State University
Ames, IA

Mark L. Tamplin
Professor in Food Microbiology
Director, Food Safety Centre
School of Agricultural Science/Tasmanian
Institute of Agricultural Research
University of Tasmania, Private Bag
54, Room 320 Life Sciences Building Sandy Bay, TAS 7005, Australia

Daniel Voit
Blentech Corporation
Santa Rosa, CA

Preface

This reference book is based on the content of the Ohio State University's Annual Thermal Processing of Ready-to-Eat Meat (RTE) Products Short Course. Thermal processing relates to heating and cooling of meat products and should not be confused with the retort process, which has traditionally been referred to as "thermal processing." This course was established to provide the latest information to the meat industry regarding all aspects of thermal processing of meat products to produce RTE products, including lethality of pathogens during the cooking process, chilling of cooked products to prevent outgrowth of spore-forming pathogens, validating the effectiveness of the heating and chilling processes, and regulatory background and requirements related to RTE meat products. Over time, sanitation and nonthermal intervention process and ingredient presentations have been added to the thermal processing focus of the original program. Food scientists are updated on food microbiology and food engineering concepts that they need to safely process RTE meat products. This first edition is the result of ten years of accumulated expertise related to the thermal processing of RTE meat products and it offers a unique compilation of technology from multiple disciplines.

This book begins with chapters that present basic heat and mass transfer, as well as microbiology information that should prepare the reader for subsequent chapters. Applications of the heat transfer and microbiology technology are made in later chapters that address batch and continuous thermal processing in air and by indirect heat transfer. Use of predictive equations, pathogen lethality, and growth models, as well as sources of supporting documentation materials for HACCP decisions, are explained. Although not thermal processing concepts, the use of various antimicrobial agents and processes, as well as sanitation principles and third-party audits are included in this book. Readers are challenged to think beyond the minimal requirements and guidelines for food safety in the final chapter.

THERMAL PROCESSING OF READY-TO-EAT MEAT PRODUCTS

CHAPTER 1

Heat and Mass Transfer

Bradley P. Marks, Michigan State University

Introduction

Thermal processing of ready-to-eat (RTE) meat and poultry products encompasses a wide variety of product categories, processing objectives, and equipment types. Obviously, developing new products and processes, or improving existing ones, requires specialized knowledge. However, if that knowledge is limited to application-specific experience, then opportunities to improve processes, ensure product safety, achieve quality objectives, and maximize profitability will be similarly limited.

The good news is that even though there is wide diversity in cooking systems, in terms of design and operation, they all operate according to the same fundamental physical principles. These principles, known as various laws of heat and mass transfer, are the subject of this chapter. To be clear, it is *not* the goal of this chapter to make the reader an expert in heat and mass transfer or computational engineering tools, for which entire books have been written. Rather, this chapter is directed expressly at individuals involved in the development, operation, and improvement of thermal processes for RTE meat and poultry products.

Specifically, it is the goal of this chapter to enable the reader to evaluate new or existing cooking systems and processes based on fundamental principles, rather than solely on prior experience. In doing so, the reader should be better equipped to evaluate how process or product modifications will impact the critical outcomes, such as end point temperatures, cooking times, or cooking yields.

Terminology and Definitions

Prior to discussing the mechanisms of heat and mass transfer, it is important to establish the critical terminology and units used in this area. Therefore, some of the most important terms are defined below.

Temperature

Three different temperatures—dry bulb, wet bulb, and dew point—are defined below. Each can be reported in standard U.S. units (Fahrenheit, °F) or International System (SI) units (Celsius, °C). In certain special cases, such as calculation of thermal radiation, the absolute temperature scales (Rankin or Kelvin, represented by °R and K, respectively) are used.

Dry-bulb temperature is a measure of the average kinetic molecular energy of matter. Practically speaking, it is the temperature (of the air or of a product) that is measured when using a dry thermometer or temperature probe.

Wet-bulb temperature is the temperature measured when the measuring point of a thermometer or temperature probe is covered by a continuously wet sock and exposed to moving air. The evaporation of water from the sock lowers the temperature of the thermometer. The number of degrees lowered, from the dry-bulb temperature, depends on the humidity of the air; at a specific dry-bulb temperature, lower air humidity results in a lower wet-bulb temperature.

Dew point temperature is also known as the saturation temperature. If moist air is cooled, this is the temperature at which condensation will begin to occur. In practical terms, if the surface temperature of a meat product or an exposed pipe or an exposed ceiling surface is below the dew point temperature of the air in contact with that surface, water vapor will condense onto the cool surface. In essence, therefore, dew point temperature is actually a measure of air humidity, which will be discussed below. All three of the temperatures defined here, and air humidity and energy, are linked by thermodynamic principles, commonly referred to as psychrometrics of moist air.

Energy and Power

Increasing or decreasing the temperature of a product requires the addition or removal of thermal energy. The relevant units of energy are as follows:

- The *British thermal unit* (Btu) is the amount of energy required to raise the temperature of 1 lb of water 1°F.
- The *calorie* (cal) is the amount of energy required to raise the temperature of 1 g of water 1°F.
- The *Joule* (J) is the SI unit for energy. It is equivalent to approximately 0.00095 Btu or 0.24 cal.

Power is the rate of energy addition or removal. In U.S. units, thermal power is typically expressed as Btu/h. In SI units, it is typically expressed as watts (W), which are equivalent to J/s. One watt is approximately 3.4 Btu/h. In relating typical mechanical power units to thermal power, one horsepower (hp) is approximately 2544 Btu/h or 745.7 W.

Humidity

In cooking systems and process environments, control of air humidity is an extremely important factor. However, humidity level is often expressed in different scales by different equipment manufacturers; so, it is extremely important to understand the difference between the scales and to know which scale is being used when describing equipment performance or process conditions. The *absolute humidity* scales are those that express the amount of water vapor in air independent of the air temperature; these scales are moisture by volume (MV), water vapor pressure (p_{vap}), humidity ratio (H or W), and dew point temperature (T_{dp}). In contrast, the relative humidity (RH) scale depends on the air temperature, as explained below.

Absolute Humidity Scales

MV describes the fraction of moist air volume that is taken up by water vapor, on a scale of 0–100%. Therefore, perfectly dry air has an MV of 0%, and pure steam has an MV of 100%. Figure 1.1 shows that the maximum possible MV is less than 100% at dry-bulb temperatures less than 212°F (100°C), because pure steam is not possible at atmospheric pressure and temperatures less than 212°F. *Therefore, MV is particularly well suited for quantifying humidity in processes operating above 212°F (100°C).*

p_{vap} describes the partial pressure of water vapor in moist air (a mixture of dry air and water vapor). At atmospheric pressure, this scale ranges from 0 to 1 atm (approximately 14.7 psi or 101 kPa). Again, perfectly dry air has a p_{vap} of 0 atm, and pure, saturated steam at 212°F (100°C) has a p_{vap} of 1 atm.

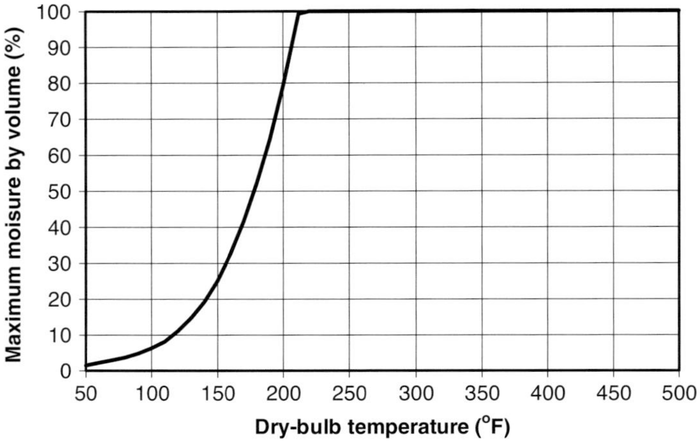

Figure 1.1. The maximum possible moisture by volume (MV) versus dry-bulb temperature for moist air at atmospheric pressure.

H quantifies the mass of water vapor per mass of dry air. Perfectly dry air has an H of 0 (lb water vapor)/(lb dry air), and pure steam has an H of infinity (lb water vapor)/(lb dry air). As such, this is not a particularly common or useful scale when describing high humidity cooking conditions.

T_{dp} is defined above. However, as noted, dew point is actually a measure of humidity that is independent of dry-bulb temperature. When given alone, the T_{dp} gives an absolute measure of humidity for moist air at atmospheric pressure, with a maximum value of 212°F (100°C). Table 1.1 compares equivalent humidity values across the scales for MV, H, and T_{dp}.

Table 1.1. A comparison of equivalent humidity values on three different absolute humidity scales

Moisture by Volume (%MV)	*Absolute Humidity (lb Water)/(lb Dry Air)*	*Dew Point (°F)*
1	0.00628	45
5	0.0327	91
10	0.0691	116
20	0.155	141
40	0.415	168
80	2.49	201
100	Infinity	212

Adapted from Machine Applications Corporation (1999).

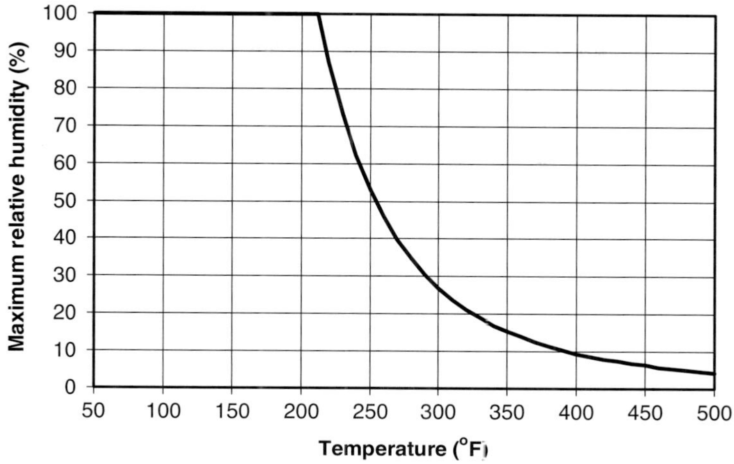

Figure 1.2. The maximum possible relative humidity (RH) versus dry-bulb temperature for moist air at atmospheric pressure.

Relative Humidity

In contrast to the absolute humidity scales, the RH of moist air depends on the dry-bulb temperature. In common terms, RH is the percent saturation of moist air (0–100%). By definition, RH is the ratio of the p_{vap} in the moist air to the saturation vapor pressure at the relevant temperature, such that $RH = 100 \times (p_{vap})/(p_{sat})$. Because p_{sat} is a function of temperature, RH is temperature-dependent. Therefore, although RH is a convenient scale (0–100%), RH alone does not quantify the absolute amount of water vapor in a moist air environment; a temperature is also needed. Additionally, Fig. 1.2 shows that the maximum possible RH is less than 100% at dry-bulb temperatures above 212°F (100°C), because the saturation vapor pressure exceeds atmospheric pressure at those conditions. *Therefore, RH is a useful humidity scale for process conditions less than 212°F, but is not well suited for conditions above 212°F, where the actual scale is no longer 0–100%.* Consequently, as noted above, it is extremely important to be aware of which humidity scale is being used when describing or evaluating oven cooking systems, and to use the best scale for a given process.

Overarching Principles

Before specifically discussing either heat or mass transfer, it is useful to present an overall concept for the flow of heat and mass. Consider the analogy that "water flows downhill"; we can see that the rate of water flow increases with greater change in elevation (hydrostatic head) and decreases with greater resistance to flow (such as a decreasing pipe diameter). The same general concept can be applied to almost any type of flow, whether electricity, heat energy, or water or oil moving in or out of a meat product. Therefore, we can apply the following general expression to all of these cases:

$$\text{flow} = \frac{\text{"driving force"}}{\text{"resistance"}}$$

In the case of heat transfer, the driving force that causes energy flow is a difference in temperature, with heat flowing "downhill" from regions of higher to regions of lower temperatures. For example, heat flows from hot oven air onto a cool product surface. Similarly, during product chilling, heat flows from the hot center of a cooked product outward to the lower temperature surface. In each case, the resistance to heat flow is related to the properties of the food product and the nature of the process conditions, which is described in more detail below.

In the case of mass transfer, the driving force is a difference in concentrations, with mass (e.g., water or oil) flowing "downhill" from regions of higher concentration to regions of lower concentration. For example, in a dry oven, water flows from the center of a meat product, which is at higher moisture content, toward the product surface, which is at lower moisture content. Likewise, in an immersion fryer, oil flows into the product, from the point of higher concentration (at the product surface) toward the interior of the product, which has a lower oil concentration.

This general principle forms the foundation upon which the following sections describe the specific mechanisms of heat and mass transfer, and the impact of product and process characteristics on heat and mass flow.

Heat Transfer

There are three basic modes of heat transfer: *conduction*, *convection*, and *radiation*. Although condensation is often referred to as a mechanism of heat transfer, it is actually a mass transfer phenomenon (even though it involves a significant energy exchange); therefore, it is discussed in the subsequent section on mass transfer. Additionally, although the following descriptions of heat transfer mechanisms present each as an independent phenomenon, the process of meat cooking actually involves complex interactions between multiple modes of heat transfer and mass transfer.

Therefore, a complete analysis of heat and mass transfer during thermal processing of meat and poultry products requires a much more rigorous approach than can be presented in this overview. Nevertheless, the goal here is to establish a basic understanding of the principles that govern the movement of heat and mass during thermal processes.

Mechanisms of Heat Transfer

Conduction

Conduction is a molecular-level mechanism by which heat energy moves through a mass (Fig. 1.3). Because molecules at higher temperature have greater energy (via vibrations), they transmit that energy via interactions with neighboring molecules that are at a lower energy level. The overall rate of heat conduction is described by Fourier's law:

$$q = -k \times \frac{\Delta T}{\Delta x} \tag{1.1}$$

where q is the heat flux, or heat flow per area (Btu/h/ft^2 or W/m^2), k is the thermal conductivity of the material through which conduction occurs (Btu/h/ft/°F or W/m/°C), ΔT is the temperature difference, and Δx is the material thickness of interest. If we refer to the previous concept that flow = "driving force" ÷ "resistance," then Fourier's law shows that the driving force for conduction is ΔT, and the resistance is $\Delta x/k$, which is also known as the R value. Therefore, this expression supports the intuitive conclusion that a thicker meat product has a greater resistance to heat flow.

The thermal conductivity, k, then is a fundamental property of the material, which depends on the chemical composition and structure of the material. For comparison, the approximate thermal conductivities of

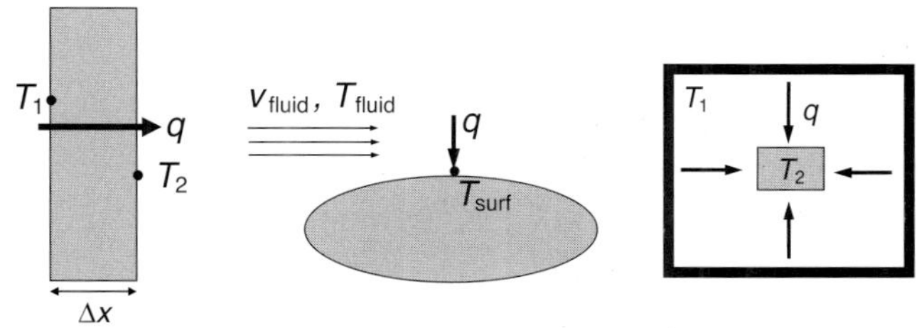

Figure 1.3. Conceptual illustrations of the three modes of heat transfer: (a) conduction, (b) convection, and (c) radiation.

aluminum, stainless steel, plywood, and fiber insulation board are 220, 16, 0.12, and 0.049 W/m/°C, respectively (Perry and Green, 1997). The k value of unfrozen meat products is typically 0.2–0.5 W/m/°C, showing that meat products are not especially good heat conductors. Water is the most important factor influencing k, which decreases with decreasing moisture content.

Convection

Convection is the movement of heat energy from a fluid to a surface or from a surface to a fluid, due to flow of the fluid (Fig. 1.3). This phenomenon is described by Newton's law of cooling:

$$q = h \times (T_{\text{fluid}} - T_{\text{surface}}) \tag{1.2}$$

where q is again the heat flux (Btu/h/ft^2 or W/m^2), h is the convective heat transfer coefficient (Btu/h/ft^2/°F or W/m^2/°C), T_{fluid} is the temperature of the bulk fluid medium (e.g., air or water or oil), and T_{surface} is the surface temperature of, for example, the meat product. Based on our general concept, this shows that the driving force for heat convection is the difference in temperature between the bulk fluid and the surface; a positive value results in heat flow into the surface, and a negative value results in heat flow out of the surface. The resistance to convective heat flux is then $1/h$.

The convection coefficient (h) is a function of the fluid properties and the fluid velocity. For example, the h values for natural air movement, forced air from a circulating room fan, and an impingement jet in a commercial oven are approximately 10, 30, and 100 W/m^2/°C. In contrast, the h value for moving water is on the order of 1,000 W/m^2/°C, showing that water is much more effective than air for convective heat transfer.

Radiation

The last mechanism of heat transfer, thermal radiation, is quite different from conduction and convection. In radiation, heat energy moves directly from one object to another via electromagnetic waves (Fig. 1.3), and therefore requires no direct molecular contact or transfer medium. The Stefan–Bolzmann law describes the rate of radiation heat transfer for a small object completely enclosed inside another object (e.g., a meat product inside an infrared oven):

$$q = \varepsilon_2 \times \sigma \times \left(T_2^4 - T_1^4\right) \tag{1.3}$$

where q is again the heat flux (Btu/h/ft^2 or W/m^2), ε_2 is the emissivity of the small object (dimensionless, 0–1), σ is the Stefan–Boltzmann constant (5.676×10^{-8} W/m^2/K^4), and T_2 and T_1 are the absolute temperatures (K or °R) of the small object and enclosure, respectively.

The emissivity describes the fraction of incident energy that is absorbed by the object, and this value happens to be quite high (~0.9) for meat products. The driving force for radiation heat transfer can then be considered as the difference between the two temperatures raised to the fourth power, so that thermal radiation becomes significant when one of the surfaces (e.g., the inside surface of an infrared oven) is very high. The resistance is represented by $1/\varepsilon$, so that a higher ε corresponds to less resistance and greater radiative heat flux.

Relative Importance of Heat Transfer Mechanisms to Cooking

As noted at the outset, thermal processing of meat products can involve all three heat transfer mechanisms: conduction, convection, and radiation. The degree to which each mechanism influences the process outcome depends on the product and process characteristics. For example, when cooking a meat patty in an impingement oven, the heat flux caused by radiation from the oven walls (hot metal) to the product is typically more than an order of magnitude less than the heat flux caused by convection from the hot air to the product surface, so that thermal radiation can be neglected in analyzing that process.

For that same product/process, a comparison of the thermal resistance due to convection ($1/h$, outside the surface) to the thermal resistance due to conduction ($\Delta x/k$, inside the product) reveals (calculations not shown) that $R_{\text{convection}} \approx R_{\text{conduction}}$. The value in this analysis is to illustrate that the product and process are well matched. If, for example, $R_{\text{convection}}$ is much less than $R_{\text{conduction}}$ for a given product and process, then fan energy is being "wasted," because the internal conductive resistance is primarily controlling the rate of cooking, and the low convective resistance indicates that the air velocity could be less without significantly reducing the rate of cooking. Intuitively, this analysis also holds true for cooking large meat products (e.g., a whole ham), which have much higher conductive resistances (due to larger dimensions) and are therefore typically cooked in ovens with much lower air velocity (thereby matching the external and internal resistances to heat flow).

Although it is beyond the scope of this chapter, it is worth noting that analytical methods exist to relate product and process conditions to product temperature and cooking time (e.g., How long will it take to raise the core temperature of a hot dog to 160°F in a water cooker?). In practice, the simplifying assumptions in these analytical methods make it very difficult to generate highly accurate predictions of process outcomes; therefore, more complex, numerical methods are necessary to generate accurate predictions of absolute outcomes. Nevertheless, simplified heat transfer analyses can be quite useful in predicting the *relative* change in process outcomes for specific changes in product or process characteristics. For example, if the required cooking time is known for a given product and process, it is possible to reasonably estimate the impact of doubling the product diameter on required cooking time (at least sufficiently well for rough estimates of changes in capacity). However, the final caution is that most of these analyses presume that only heat transfer occurs, without any concurrent mass transfer, which is the topic of the next section.

Mass Transfer

Conceptually (and mathematically), mass transfer is identical to heat transfer (with a few important differences). The driving force for mass flow between two points can be caused by multiple mechanisms, including gravity, capillary action, osmosis, and concentration differences. In reality, all the above can occur within a meat product subjected to a thermal process. However, for simplicity, we consider only the two primary, concentration-driven mechanisms: diffusion (analogous to heat conduction) and mass convection (analogous, of course, to heat convection).

Mechanisms of Mass Transfer

Diffusion

Mathematically identical to heat conduction, diffusion is the process by which molecules of one substance (e.g., water) move through another substance (e.g., a meat product), from regions of higher concentration to regions of lower concentration. This mechanism is described by Fick's law:

$$n = -D_{\mathrm{AB}} \times \frac{\Delta c_{\mathrm{A}}}{\Delta x} \tag{1.4}$$

where n is the flux of substance A (lb/h/ft^2 or kg/s/m^2), D_{AB} is the mass diffusivity of substance A through substance B (e.g., water in whole-muscle meat), Δc_{A} is the concentration of substance A in B (e.g., the dry basis moisture content of a meat product), and Δx is the material thickness of interest. Again, we can see that the driving force for diffusion is the concentration difference, and the resistance is $\Delta x / D_{AB}$.

Unlike thermal conductivity (k), mass diffusivity (D_{AB}) is a much more difficult value to measure or locate; however, some values can be found in the literature for meat products. The noteworthy characteristic of D_{AB} is that it is heavily influenced by the moisture content of the product. As moisture content decreases, particularly if dealing with an intermediate- or low-moisture product, D_{AB} decreases significantly. The result is that the resistance to mass flow increases with decreasing moisture content.

Convection

Mass convection is analogous and fundamentally related to heat convection, such that mass convection is described by:

$$n = h_m \times (c_{A,\text{fluid}} - c_{A,\text{surface}}) \tag{1.5}$$

where n is again the flux of substance A (lb/h/ft^2 or kg/s/m^2) to or from the surface, h_m is the convective mass transfer coefficient (ft/min or m/s), $c_{A,\text{fluid}}$ is the concentration of A in the bulk fluid (e.g., MV of the air), and $c_{A,\text{surface}}$ is the concentration of A in the fluid right at the product surface. It should be clear by now (hopefully) that the driving force for mass convection is the difference between the bulk and surface concentrations and that the resistance to mass convection is expressed by $1/h_m$. The convective mass transfer coefficient (h_m) is directly related to the convective heat transfer coefficient (h), with the relationship governed by the properties of the fluid being considered. Regardless of the process, h_m is related to fluid velocity in the same way as is h; increasing fluid velocity results in increased h_m and, therefore, more effective mass convection.

A key difference between heat and mass convection is in the value of $c_{A,\text{surface}}$. In heat transfer, the temperature of the product surface is equal to the temperature of the air right at the product surface. However, in mass transfer, the concentration in the solid at the surface is not equal to the concentration in the fluid right at the surface. The two values ($c_{A,\text{surface,product}}$ and $c_{A,\text{surface,fluid}}$) are directly related, but are *not* equal. In other words, at an equilibrium condition, the moisture content at the product surface is directly linked to the air humidity and temperature, but it is not *equal* to the concentration of water in the air. The relationship between the two can be found in the literature for some meat products.

Condensation and Evaporation

Given the above, condensation and evaporation are just special (and very important) cases of mass convection to and from moist air. Condensation occurs when $c_{\text{water,air,bulk}} > c_{\text{water,air,surface}}$, and evaporation occurs when $c_{\text{water,air,bulk}} < c_{\text{water,air,surface}}$. In both cases, $c_{\text{water,air,bulk}}$ is the absolute humidity of the process air. In the case of condensation, $c_{\text{water,air,surface}}$ is the saturation humidity at the surface temperature; in the case of evaporation, $c_{\text{water,air,surface}}$ is the equilibrium absolute humidity corresponding to the current surface moisture content of the product. In different, and more measurable, terms, condensation occurs when $T_{\text{air,dew point}} > T_{\text{product,surface}}$, and evaporation occurs when $T_{\text{air,dew point}} < T_{\text{product,surface}}$.

When water condenses or evaporates at a product surface, a significant amount of energy is also involved. That energy, called the latent heat of vaporization (λ_{v}), is the amount of energy necessary to change water from a liquid to a gas ($\sim$970 Btu/lb or $\sim$2200 kJ/kg). As a comparison, this is approximately five times more energy than it takes to increase the temperature of liquid water from 32 to 212°F (0 to 100°C); therefore, condensation and evaporation play a very important role in the energy transfer during cooking. The total flux during cooking of a meat product in a moist air process can be expressed as:

$$q = h \times (T_{\text{air}} - T_{\text{surface}}) + \lambda_{\text{v}} \times h_{\text{m}} \times (c_{\text{air,bulk}} - c_{\text{air,surface}}) \qquad (1.6)$$

where all the variables have been previously defined. The first term on the right side of the equation shows the heat convection contribution to total heat flux. The second term on the right shows the condensation/evaporation (mass convection) contribution to heat flux. That term shows that the energy gained or lost due to condensation or evaporation is the rate of mass convection times the latent heat of vaporization. This is a particularly important point to make—that condensation and evaporation can contribute significantly (positively and negatively, respectively) to the net heat flux to a product in an oven. In the equation above, it can also be noted that the resistances ($1/h$ and $1/h_{\text{m}}$) will be relatively constant down an oven line (because they are controlled by the air flow conditions), but the driving forces (ΔT and Δc) will both vary with time of cooking, as the product surface temperature and moisture content change.

Summary

This chapter has outlined the underlining mechanisms by which heat energy is transferred into or out of meat products during cooking or chilling. Although rigorous, engineering analyses of these mechanisms are beyond the scope of this chapter, an understanding of these basic principles is important to anyone involved in production of RTE meat and poultry products. Although the jargon associated with cooking systems can vary sector to sector and vendor to vendor, the basic physical principles governing heat and mass transfer control the outcomes of any cooking process, and personnel who understand these principles will be better equipped to evaluate and improve thermal processes.

References

Perry, R.H. and Green, D.W. 1997. *Perry's Chemical Engineers' Handbook*, 7th edn. McGraw-Hill, New York.

Machine Applications Corporation. 1999. *The MAC Humidity/Moisture Handbook*. Machine Instruments Corporation, Sandusky, OH. Accessed at: http://www.macinstruments.com/pdf/handbook.pdf.

CHAPTER 2

Microbiology of Cooked Meats

Aubrey F. Mendonca, Iowa State University

Introduction

Microorganisms on raw red meats and poultry include mesophilic and psychrotrophic bacteria, and yeasts and molds from the hide and intestinal tract of the animal itself, the animal's environment, and from various sources in slaughter and meatprocessing facilities (Jackson et al., 2001). Even though the cooking of raw red meats and poultry can drastically reduce initial numbers of microorganisms, bacterial spores and some thermoduric bacteria such as certain micrococci, enterococci, and lactobacilli can survive. Postcook handling of meats in refrigerated meat processing environments will contribute various populations of psychrotrophic bacteria to those products depending on the sanitary condition of the processing facility. Contamination of cooked meats with psychrotrophic bacteria is important because of the ability of psychrotrophs to grow and spoil food products even at proper refrigeration temperatures. In this regard the higher the initial numbers of psychrotrophic bacteria that contaminate cooked meats, the shorter will be the shelf life of those products during refrigerated storage. In addition, the growth of psychrotrophic pathogens during extended refrigerated storage of cooked meat products is a major food safety concern.

The growth of microorganisms, including bacteria, yeast, and molds, is of great importance to meat processors as it is to all food processors. For example, the controlled growth of lactic acid producing bacteria, including certain *Lactobacillus* spp. and *Pediococcus* spp., is very important in the production of fermented meat products. In contrast, uncontrolled growth of bacteria, yeasts, and molds can cause spoilage as well as compromise the safety of meat products. The growth of yeasts, which are usually present in low numbers in meats, is important in meat spoilage especially when the meat surface becomes dry. Uncontrolled growth of microorganisms can ruin the quality of a few products or an entire batch of products and cause huge financial losses to a company. Even though the cooking process can drastically reduce initial numbers of microorganisms, bacterial spores and some thermoduric vegetative cells can survive. Therefore, it is very important to prevent microbial contamination of the finished product and to ensure that the product is subsequently stored at a temperature that is low enough to prevent microbial growth. The microbial safety and quality of cooked meats and the success of a small meat processor or large meat processing company depend largely on the microbial quality of the raw materials, the cooking process, preventing postcook contamination, and effective control of microbial growth during storage and distribution of the finished product.

Sources of Microorganisms in Raw Meat

During slaughter of meat animals and the carcass dressing process, many types of microorganisms, mostly bacteria, contaminate the carcasses. These microbial contaminants come from the pasture and/or feedlot environment (soil, dust, water, feed, and manure), the animals themselves (skin, hide, hair, feathers, and gastrointestinal tract), and the slaughtering facilities (air, aerosols, knives, equipment, water, and workers). In the process of dehiding the slaughtered cattle, microorganisms are transferred from the hide to the carcass via knives and other cutting implements, and the plant workers. Accidental spillage of intestinal contents during evisceration results in gross contamination of the carcass. Following slaughter, dressing, and washing, the numbers of aerobic bacteria on hog carcasses range from about 10^2 to 10^4 per cm^2 (Roberts et al., 1980); numbers on cattle carcasses range from about 10^3 to 10^5 per cm^2 (Ingram and Roberts, 1976). Among the microbial contaminants on animal carcass, relatively low numbers of enteric pathogens such as *Salmonella* spp., *Campylobacter jejuni*, *Escherichia coli*, *Staphylococcus aureus*, and *Clostridium perfringens* may be present. Poultry carcasses, as compared to carcasses of other food animals, generally carry higher numbers of *Salmonella* from fecal contamination. The spread of microorganisms on meat continues during preparation of primal and subprimal cuts and during grinding, mixing, and comminuting.

Effect of Cooking on Microorganisms in Meat

The microbiology of cooked meats is influenced by the initial microbial populations of the raw materials and the cooking process. The process of cooking, depending on the time and temperature, can destroy many microorganisms in meat. However, bacterial spores and thermoduric vegetative cells survive. The number of bacterial cells that survive cooking of meat, for example, at 150°F (65.5°C) is proportional to the numbers initially present. Therefore, it is a serious mistake to mishandle meats with the belief that "cooking will destroy the higher microbial load anyway." The chances of sublethally injured microorganisms surviving in cooked meats are greater when raw materials of poor microbial quality are used and the heat treatment is inadequate. The presence of sublethally injured bacteria in cooked meats is an insidious problem because these organisms may be able to repair their injury over time and, if conditions are favorable, grow to cause spoilage or pose a public health hazard. It should be noted that heat treatments used in meat processing, except for canning, are not aimed at sterilizing meat products. In fact the purpose of the cooking in the production of ready-to-eat (RTE) meat products is to provide microbiological safe products that offer flavor, appeal, and convenience.

Sources of Microorganisms in Cooked Meats

Cooked meat products that may be uncured or cured and heated to an internal temperature of approximately 160°F (71.1°C) include roasts, frankfurters, bologna, some hams, and luncheon meats. These meat products are packaged aerobically or anaerobically, and held at refrigeration temperature. Sources of microbial contaminants on these products prior to heating include the raw meat, ingredients used in formulation, air, processing equipment, and plant workers. Spices, if not sterilized, can contribute large numbers of bacterial spores, yeast, and molds to the product. Heating meat products to an internal temperature of 160°F (71.1°C) or higher will destroy most microorganisms except for thermoduric bacteria (e.g., certain *Lactobacillus*, *Micrococcus*, and *Enterococcus*) and spores of *Bacillus* and *Clostridium*. Generally, the microbial numbers in freshly prepared cooked meat products are about 10^2 or less per gram (Tompkin et al., 2001). After heating, some of these products are subjected to further processing and handling.

Products that undergo removal of casings, slicing, and packaging, may inevitably come in contact with equipment, recycled brine, air, and workers before final packaging. Depending on the cleanliness and sanitary condition of the processing environment, various types of bacteria, yeasts, and molds can contaminate cooked meat products. Even though the initial bacterial population in these products rarely exceeds 10^2 (colony forming units) CFU/g, the contaminants can be psychrotrophic facultative anaerobic and anaerobic bacteria including, *Listeria*, *Lactobacillus*, *Leuconostoc*, and certain *Clostridium* spp. These organisms can multiply to relatively high numbers during extended storage of cooked meats in vacuum or in controlled-atmosphere packages and compromise the safety and shelf life of these products. This problem is further exacerbated in low-fat meat products that have high pH and high water activity, and by fluctuation of storage temperature.

Factors Affecting Microbial Growth in Cooked Meats

The factors that affect microbial growth in cooked meat products are generally similar to the factors that affect growth of microorganisms in other foods. These factors can be placed into the following two categories: (a) factors that are associated with the meat product itself, for example, nutrient availability, water activity, pH, and presence of growth inhibitors, and (b) factors that are associated with the storage environment, for example, temperature and gaseous atmosphere.

Nutrient Availability

Cooked meats are nutrient-rich and therefore provide a more than adequate amount of protein, carbohydrate, lipids, vitamins, minerals, and other growth factors to support the growth of bacteria, yeasts, and molds. Unfortunately, there is no practical way in manipulating nutrient availability in meats to inhibit microbial growth. However, other factors can be manipulated to slow down or prevent the use of available nutrients by contaminating microorganisms.

Water Activity

Water activity (a_W) may be defined as the amount of free water that is available for microbial growth. Of the total water content of a cooked meat product, only a portion of that water may be available to microorganisms. The remainder may be chemically bound to certain components of the product such as sodium chloride, phosphates, and lactate. The higher the water activity, the more free water is available for microbial growth. Pure water has a_W of 1.00 and a 16% (w/v) sodium chloride solution has a_W of 0.90. Generally, most spoilage bacteria cannot grow below a_W of 0.91; however, molds can grow at a_W as low as 0.80. With respect to foodborne pathogens, *S. aureus* can grow at a_W as low as 0.86. In contrast, *C. botulinum* does not grow below a_W of 0.94 (Jay et al., 2005).

pH

The degree of acidity or alkalinity of food is frequently expressed in terms of pH. It is well known that most microorganisms grow best in a pH range of about 6.6–7.5. Therefore, meat at pH 6.0–6.2 will spoil faster than meat at pH 5.2–5.4. Some lactic acid bacteria, for example, *Lactobacillus brevis* and *L. plantarum*, can grow at pH 3.16 and 3.34, respectively. Generally, at pH below 4.0 yeasts grow better than bacteria, whereas only molds can be found at pH values below 1.5 (Jay et al., 2005).

The acidification of meats is the preservative principle involved in sausage fermentation. The fermentation of summer sausage from the intentional addition of lactic acid bacteria (*Lactobacillus* spp., *Pediococcus* spp., and/or *Micrococcus* spp.) results in the low pH of summer sausage (pH 4.7–5.0). This relatively low pH inhibits the growth of foodborne pathogens including toxin-producing *S. aureus* (Jay et al., 2005). It should be noted that although lactic acid bacteria are beneficial in summer sausage, they are considered as spoilage organisms in vacuum-packaged fresh or cooked meats.

Presence of Growth Inhibitors

Certain components of RTE meats such as sodium chloride, nitrite, polyphosphates, lactates, diacetates, spices, and constituents of smoke can inhibit microbial growth. A sodium chloride concentration of 5% (w/v) inhibits the growth of many Gram-negative spoilage organisms including *Pseudomonas*. Salt tolerant microbial types, for example, *Micrococcus* and *S. aureus*, are not inhibited. Sodium nitrite is an effective inhibitor of *C. botulinum* growth and toxin production in vacuum-packaged cured meats. Some polyphosphates may offer limited safety margin against *C. botulinum* in cooked vacuum-packaged uncured meats. Some spices used in sausage formulations contain natural antimicrobial compounds. Also, certain constituents of smoke are known to have bacteriostatic or bactericidal properties.

Storage Temperature

Storage temperature is one of the most important factors that influences microbial growth and thus spoilage of perishable foods. Microbial growth takes place via the action of enzymes. It is well established that with every 10°C rise in temperature within the reaction range, the rate of enzymatic reactions doubles. Conversely, by decreasing the temperature by 10°C the rate of enzymatic reaction is reduced by half. For long-term storage, meats need to be stored at 41°F (5°C, refrigeration) to −4°F (−20°C, freezing) or lower.

Gaseous Atmosphere

Microbial growth is influenced by the composition of gaseous environment. For example, the rapid growing aerobic spoilage organisms on meat are inhibited when meat is vacuum-packaged or packaged under modified atmosphere. Modified atmosphere packaging of meats involves packaging these products in atmospheres that contain increasing amounts of carbon dioxide (CO_2) up to 10%. CO_2 is very inhibitory to aerobic organisms. Microbial inhibition by CO_2 increases with decreasing temperatures and is mainly associated with increased solubility of the gas at lower temperatures. In addition, the pH of meats stored in high CO_2 environments was found to be slightly lower than that of controls, which were stored in air, due to the formation of carbonic acid. Gram-negative bacteria demonstrate greater sensitivity to CO_2 than Gram-positive bacteria, with *Pseudonomas* spp. being highly sensitive and the lactic acid bacteria and anaerobes being very resistant (Jay et al., 2005).

Microbial Spoilage of Cooked Meats

The main spoilage defects caused by microbial growth in cooked meats are slime on the surface of the product, souring, and discoloration. Slime on meat products is produced by the growth of bacteria or yeast. Millions of these microorganisms grow to form tiny colonies, which eventually enlarge and join together to form a mass of gray slime. Slime usually occurs on the outer surface of the meat product and is readily formed when the surface is moist. Souring is caused by lactic acid producing bacteria such as lactobacilli, enterococci, and related organisms that usually come for the raw meat or from milk solids added to meats during processing. These organisms produce acids during their growth and use of lactose or other sugars. Souring generally occurs underneath the casing of cooked meat products (Jay et al., 2005). Discoloration in cooked meats can be attributed to chemical as well as microbiological factors. For example, two types of greening can occur in processed red meats. One type that is caused by hydrogen peroxide (H_2O_2) commonly occurs in frankfurters and other cured red meats that are vacuum packaged. Upon exposure of these meat products to air, a small amount of H_2O_2 may form and react with the cured meat pigment, nitrosohemochrome, to produce a greenish color. Greening also occurs when certain bacteria in the interior core of the cured meat product grow and produce H_2O_2 (Pearson and Gillett, 1999). In such instances, discoloration is seen as green rings or green cores in the meat product. Usually, high levels of rework that lead to increased survival of microbial contaminants during cooking contribute to discoloration in meat products. Table 2.1 shows a summary of some microbial spoilage defects in vacuum-packaged or modified atmosphere packaged cooked meat products.

Perishable Cooked Uncured Meats

Most cooked uncured pork and poultry products are heated to relatively high temperatures resulting in destruction of vegetative cells but not bacterial spores. Beef products are usually heated to lower temperatures that destroy nonsporeforming pathogens; however, some thermoduric organisms survive to contribute to the spoilage microflora of these products. Uncured meat products, such as beef and turkey roasts, are heated to an internal temperature ranging from 140 to 150°F (60–65°C). Depending on the size of the roast, which could be 10 lb or more, the product surface

Table 2.1. Summary of some microbial spoilage defects in vacuum-packaged or modified atmosphere packaged cooked meat products

Product	*Spoilage Defect*	*Organism*	*Reference**
Vacuum-packaged uncured turkey breast meat	H_2S odor and gas	Psychrotrophic clostridia	(a)
Vacuum-packaged roast beef	H_2S odor and gas	Psychrotrophic clostridia	(a)
Vacuum-packaged bologna	Greening	*Carnobacterium viridans*	(b)
Frankfurters packaged in modified atmosphere (CO_2 and N_2)	Greening	*Weissella viridescens*	(c)
Vacuum-packaged wieners and bologna	Greening/slime	*W. viridescens*	(d)
Vacuum-packaged luncheon meat	Souring	Lactic acid bacteria	(e)
Vacuum-packaged luncheon meat	Yellowing	*Enterococcus casseliflavus*	(f)

*(a) Kalinowski and Tompkin (1999); (b) Holley et al. (2002); (c) Blickstad and Molin (1983); (d) Pearson and Gillett (1999); (e) Kempton and Bobier (1970); (f) Whiteley and D'Souza (1989).

could be exposed to the final temperature for 1 hour or longer. Freshly made cooked uncured meat products will usually have 35°C plate counts of 10^2 or less per gram (Tompkin et al., 2001). Microbial survivors of such a heat treatment include some very thermoduric species (*Enterococcus*, *Micrococcus*, and *W. viridescens*) and spores of *Bacillus* and *Clostridium* spp. In addition to these microorganisms that survive the heating process in the roasts, other microorganisms can enter these products during handling prior to vacuum-packaging and subsequent storage at refrigeration temperature.

The presence of *E. coli* in cooked uncured meat products is indicative of poor sanitary conditions. In this situation it is important to investigate the meat processing facility to find the source of *E. coli* and recommend corrective action (Tompkin et al., 2001). Equipment, workers, and air are major sources of these postheating microbial contaminants on roasts. Slicing of the cooked roasts increases the surface area of these products and the likelihood of microbial contamination from equipment, workers, and the environment. Also, microorganisms that survive the cooking process are spread over the surfaces of the sliced product during cutting. The microbial load on roasts can increase further if spices, herbs, or other ingredients are added to the products after heating.

Certain characteristics of cooked uncured meat products such as their high nutrient content, favorable pH, and low salt content make these products an ideal medium for microbial growth. Microbial growth in these products will occur rapidly if they are held at favorable temperatures for extended time periods. Based on most regulations, holding precooked meats at temperatures between 41°F (5°C) and 140°F (60°C) is prohibited except during preparation, heating, or chilling. Many cooked uncured meats are frozen for wholesale distribution and if they do not thaw during commercial shipment, the microbial populations at retail will generally correspond to those present just after the products were frozen (Tompkin et al., 2001). If these products are held above freezing point for extended time periods they will spoil from a variety of psychrotrophic bacteria and yeasts. The types of psychrotrophic spoilage microflora on cooked uncured meats will be affected by factors including packaging and temperature.

Spoilage of cooked uncured meat products is caused by psychrotrophic facultative anaerobic and anaerobic bacteria. Heterofermentative *Lactobacillus* spp. and *Leuconostoc* spp. produce large amounts of gas (CO_2) and purge (from acid production) in the packaged meat, without creating major changes in color, flavor, or texture. Spoilage of certain cooked meats such as turkey breast, roast beef, and cooked pork, has been attributed to the growth of more that one type of psychrotrophic clostridia (Kalinowski and Tompkin, 1999). These meat products are cooked in heat-resistant plastic bags and stored under normal refrigeration conditions without being temperature-abused. The spoilage is of the proteolytic type and may remain undetected until removal of the plastic package. Tests performed on these products to detect toxin have yielded negative results. It is likely that the source of the clostridia is the raw meat or poultry, where similar spoilage has been observed (Broda et al., 1998a, 1998b; Kalchayanand et al., 1993).

Psychrotrophic clostridia can produce gas and purge along with off-flavors and changes in meat color, from brown to pink, to red after four weeks. *Proteus* and *Hafnia* spp. have been implicated in the spoilage of sliced roast beef, which changed from brown to pink in one week and developed a putrid smell after six weeks. Unpacked cooked meat products that do not contain carbohydrates can develop putrid odors resulting from bacterial growth and protein degradation (Tompkin, 1986; Tompkin et al., 2001).

The addition of uncooked ingredients such as spices, celery, cheese, gravies, or sauces to cooked meats will alter the microbial profile of these products so that the previously described microbial composition may not be fully pertinent. Therefore, a detailed description of product formulation and processing is necessary for meaningful interpretation of laboratory data on the microbial composition of these meat products (Tompkin et al., 2001).

Perishable Cooked Cured Meats

Cooked cured meats include wide array of products such as frankfurters, ham, bologna, and luncheon meats prepared from beef, pork, and poultry. These meat products are made from ground or chopped meats; therefore, microorganisms are distributed throughout the products and not confined only to the surface as in roasts. Additives are incorporated in these products for improving color, flavor, texture, shelf life, and microbial safety. Examples of these additives are sodium chloride, nitrite, phosphate, lactate, diacetate, dextrose, erythorbate, sorbate, soy protein, nonfat dry milk, carrageenan, and spices. Some products are low in fat ($\leq$2%) whereas others, for example some frankfurters, can have $\geq$30% fat. The pH of these products ranges from $<$6.0 to 6.8 (in some low-fat products). They are heated to an internal temperature of 150–160°F (65–71°C) and, depending on their size, the surface can be heated to the final temperature for a longer time compared to the center. Since the microorganisms are initially distributed throughout the meat mixture, thermoduric organisms that survive heating will be present throughout the final product (Tompkin et al., 2001).

The handling of these products after heating and prior to vacuum-packaging or packaging under modified atmosphere of CO_2 or CO_2 + nitrogen (N_2), exposes them to microbial contamination. Postheat handling may involve chilling, skinning, slicing, and portioning. These actions predispose meat products to microbial contamination from equipment, workers, water, and air. Frankfurters become contaminated on the surface but in sliced products the contaminating organisms are spread over the cut surfaces during slicing. Some of the microbial contaminants are established in the processing environment, particularly in areas that are difficult to clean and sanitize. Some organisms that survived the heat process and some microbial contaminants of cooked cured meats are selectively inhibited by salt, nitrite, and other antimicrobial ingredients such as sodium lactate, sodium diacetate, and buffered citric acid. Following packaging, cooked cured meat products usually have bacterial counts of about 10^3 or less per gram. Higher bacterial counts in these products from retail cases are indicative of the time–temperature history of the product (Tompkin et al., 2001).

During extended storage at refrigeration temperatures, cooked cured meats may be spoiled by lactic acid bacteria, enterococci, micrococci, yeasts, and molds. Those products that are vacuum-packaged or packaged in modified atmosphere are usually spoiled by psychrotrophic *Lactobacillus* spp. and *Leuconostoc* spp. The microorganisms that are usually involved in spoilage are homofermentative *L. sake*, *L. curvatus*, heterofermentative *W. viridescens*, and *Leuconostoc carnosum*, *L. gelidum*, and *L. mesenteriodes*. Visible signs of spoilage include gas accumulation and/or purge, and cloudy appearance (Yang and Ray, 1994). Formation of loose or gassy vacuum packages is commonly attributed to heterofermentative lactic acid bacteria. Slime from bacterial cells, dextran production by *Leuconostoc* species and an acidic flavor can occur in products that contain sucrose (Makela et al., 1992). Some microorganisms such as *Serratia liquefaciens* degrade amino acids to produce an ammonia-like odor (Ray, 2001) which can be described as a diaper smell. The growth of lactic acid bacteria in vacuum-packaged low-fat turkey rolls that are portioned and sliced can result in a pink coloration of the product. When these products are stored in oxygen-permeable packaging or unpackaged, some *Lactobacillus* species may produce hydrogen peroxide. The hydrogen peroxide oxidizes nitrosohemochrome to metmyoglobin (brown) or to oxidized porphyrins to produce a green color. Some of the bacteria that are mainly associated with greening in cured meats include *W. viridescens*, some *Lactobacillus* spp., leuconostocs, *E. faecium*, and *E. faecalis* (Jay et al., 2005). Certain bacteria such as *E. casseliflavus* can produce yellow discoloration on sliced, vacuum-packaged luncheon meats due to naturally produced pigment. *E. casseliflavus* produced discoloration in a vacuum-packaged, cooked, cured meat product during 3–4 weeks of storage at 40°F (4.4°C). This organism can survive cooking at 160°F (71.1°C) for 20 minutes but not 30 minutes (Whiteley and D'Souza, 1989).

Shelf-Stable Canned Uncured Meat Products

Typical shelf-stable canned uncured meat products include beef stew, roast beef with gravy, chili con carne, canned whole chicken, and tamales. These meat products are packaged in hermetically sealed containers (glass jars, aluminum or steel cans, semi-rigid metal foils or heat-resistant plastic pouches) and are heated in the container or aseptically packaged after heating. They are considered low acid canned foods; therefore, they must receive a minimum heat treatment designed to destroy *C. botulinum* (botulinum cook; $F_0 = 2.5$) and protected from postprocessing contamination. To control microbial spoilage in these products, heat treatments of $F_0 =$ 4–6 are required to destroy spores that have a higher heat resistance than that of *C. botulinum* spores. Also, F_0 treatments of 12–15 are necessary to destroy spores of thermophilic bacilli and clostridia in products which are destined for storage and distribution in tropical countries. Other types of shelf-stable canned uncured meat products include sloppy joe and spaghetti sauce with ground meat. These are high acid canned products which are given a milder heat-processing treatment.

Microbial spoilage of canned uncured meats can occur from inadequate thermal processing or postprocessing contamination. Inadequate thermal processing usually results in survival of spores of putrefactive anaerobes and thermophilic spoilage organisms. During storage of these products at ambient temperatures, surviving spores of the putrefactive anaerobe, *C. sporogenes* PA3679, can grow and cause spoilage. If the products are stored at temperatures above 109°F (43°C), spores of thermophilic organisms can grow and produce spoilage. Following thermal processing of canned uncured meats, microorganisms may contaminate these products though pin holes in defective or externally corroded cans, defective can seams from faulty closure or rough handling, or through faulty sealant that attaches the lid to the body of the can. During cooling of the canned product, the vacuum that develops in the container may pull cooling water into the product thus facilitating microbial contamination. Microbial spoilage of these products following thermal treatment is generally characterized by the presence of mixed cultures. Variations in the microbial composition of these mixed cultures are directly related to the microbial quality of the cooling water and the types of microorganisms endemic in the processing facility. Due to the widespread use of chlorine in cannery cooling-waters, spore formers have become the more common postprocessing contaminant of canned products (Tompkin et al., 2001).

Shelf-Stable Canned Cured Meat Products

Shelf-stable canned cured meat products can be placed into four categories based on antimicrobial procedures used to attain shelf stability. These categories are (a) products given a botulinal cook ($F_0 \geq 2.78$);

Table 2.2. Categories of shelf-stable canned cured meat products based on antimicrobial procedures used to attain shelf stability*

Product Category	*Examples*
A. Products given a botulinal cook ($F_o \geq 2.78$)	Canned frankfurters, viennas, meat spreads, corned beef, chicken or turkey luncheon meat
B. Products given less than a botulinal cook ($F_o = 0.1$–0.7)	12-oz canned luncheon meat, canned hams ($\leq$3 lb)
C. Products stabilized by water activity ($a_w \leq 0.92$)	Sliced dried beef sausages immersed in hot oil (in the final container), prefried bacon (packaged under vacuum or modified gaseous atmosphere)
D. Products stabilized by low pH (pH $\leq$ 4.6)	Meats pickled in vinegar (e.g., sausages, pigs' feet)

*Adapted from Tompkin et al. (2001).

(b) products given less than a botulinal cook ($F_o = 0.1$–0.7); (c) products stabilized by water activity ($a_W \leq 0.92$); and (d) products stabilized by low pH (pH $\leq$ 4.6). Examples of meat products in each category are shown in Table 2.2. Thermal processing along with additives such as sodium chloride, nitrite, phosphate, and lactate contribute to destruction of both spoilage and pathogenic microorganisms in these products. As with other canned foods, the heat process applied to canned meats depends on factors, including type of meat product, container size, product pH, consistency of product, and type of processing equipment (Ball and Olson, 1957). The pH of the meat product is the most important factor that establishes the degree of thermal processing needed for microbial stability of the product. The importance of pH is attributed to enhance microbial destruction during thermal processing and the inhibitory effect of acid on survival and outgrowth of microbial survivors.

The microbiology of meat products in category (a) is similar to that of low-acid canned foods. Low-acid canned foods are processed to achieve commercial sterility. Commercial sterility is defined as destruction of all *C. botulinum* spores and all other pathogenic bacteria as well as more heat-resistant organisms, which, if present, could cause spoilage under normal conditions of storage and distribution (National Canners Association, 1968). Commercially sterile foods are not absolutely sterile. Infrequently, low numbers of certain thermophilic spores may be present in canned low-acid foods including meat products; however, these organisms will only cause spoilage if the food is stored at temperatures above 109°F (43°C).

Generally, microbial spoilage of canned cured meat products in categories (a) or (b) will not occur unless these products were subjected to an improper curing process, inadequately heat treated or exposed to postprocessing microbial contamination. An improper curing process involving insufficient amounts of nitrite or salt can allow growth of sporeforming bacteria that survived the heat treatment. Inadequate heat processing will usually result in the survival of bacterial spores of *Bacillus* and *Clostridium* species. A slightly inadequate thermal process can result in survival of an unusually larger number of mesophilic spores thus increasing the likelihood that subsequent microbial growth will occur. A grossly inadequate thermal process usually results in survival and growth of enterococci and other thermoduric nonsporeformers in addition to bacterial spores. Defective or damaged containers facilitate postprocessing contamination of canned cured meat products with microorganisms from environmental sources such as cooling water, air, or dirty contact surfaces. Many groups of microorganisms may be found in products with defective or damaged containers that permit leakage. Microbial groups include cocci, coccobacilli, lactic acid bacteria (long or short rods), aerobic sporeformers, and yeast and molds (Tompkin et al., 2001).

Spoilage of meat products in category (c) including sliced dried beef, sausages, and prefried bacon will not occur unless the water activity of these products is higher than recommended or the product container is leaky due to a broken seal or damage. These products may be vacuum-packaged or packaged in modified gaseous atmosphere containing a low amount of residual oxygen. The growth of more salt-tolerant organisms such as enterococci or micrococci can occur in products with a water activity close to 0.86, especially if the products have been packaged under a low vacuum or nitrate was added during curing (Tompkin et al., 2001).

Cured meat products such as pickled pigs' feet and pickled sausage in category (d) are commonly immersed in acidified brine; therefore, the microbial stability of these products is largely attributed to low pH, acetic acid, little or no fermentable sugar in the product, and an airtight container (Niven, 1952). Aerobic plate counts are highly variable and cannot be predicted. Lactic acid bacteria and spores may be present. Cloudiness of the brine can occur when numbers of lactic acid bacteria are greater than about 10^7 CFU/g.

Perishable Canned Cured Meat Products

Perishable canned cured meat products include ham and other cured meats such as pate, pressed ham, bacon, and emulsion style sausages that contain concentrations of nitrite and salt similar to those in shelf-stable cured meats. These meat products may be cooked to approximately 149–167°F (65–75°C) in hermetically sealed high oxygen barrier films and marketed as "cook-in-bag" products. Alternatively, they may be repackaged after cooking. Unlike shelf-stable cured meats they are not heated to high temperatures that are adequate to destroy the indigenous spore populations. Therefore, they must be distributed and stored at refrigeration temperatures to ensure microbial safety and stability. When properly heated in the final container and not contaminated after heating these products usually have a shelf-life of 1–3 years at 32–41°F (0–5°C) (Tompkin et al., 2001).

Microbial spoilage of refrigerated perishable canned cured meats is influenced by various factors including characteristics of the meat product (e.g., pH, a_W, nitrite, lactate, salt), thermal process and packaging conditions, and the storage temperature (Tompkin, 1986, 1995). Spoilage at refrigeration temperatures is commonly attributed to survival and growth of psychrotrophic, thermoduric, nonsporeforming bacteria such as enterococci, *W. viridescens* and other homo- or heterofermentative lactic acid bacteria or leuconostocs. Increased survival of these spoilage organisms occurs when the raw meat ingredients have high initial microbial populations or a substandard thermal process was applied (Tompkin et al., 2001). Typical signs of spoilage include sourness or off-odor, milky exudate, slime and/or gas formation, and green discoloration. Slime in these products may be attributed to dextran produced by leuconostocs growing on sucrose or may consist of glucose and galactose produced by leuconostocs and *L. sake* (Makela et al., 1992). Green discoloration occurs when certain lactic acid bacteria such as *W. viridescens*, *L. fructivorans*, and *L. jensenii* produce hydrogen peroxide. The hydrogen peroxide oxidizes the porphyrin ring of nitrosohemochrome to form choleomyoglobin which imparts a green color to the meat product (Pearson and Gillett, 1999). Other bacteria including *E. faecium*, *E. faecalis*, and leuconostocs are capable of producing green discoloration (Jay et al., 2005). Greening can occur at the surface of the product at the center or core where there is bacterial contamination growth. Greening in the center of cured meat products has been linked to growth of *W. viridescens* (Blickstad and Molin, 1983).

Spoilage of these products at ambient temperature results from the proliferation of various thermotolerant mesophilic bacteria. Depending on the amount of oxygen that permeates the packaged meat product, the growth of *Bacillus cereus* or *B. licheniformis* on the surface of perishable cured meats can result in surface softening and off odors (Bell and De Lacy, 1983). When packaging films of low gas permeability are used spoilage is localized below the film near the clip or seam (Bell and De Lacy, 1982).

The spoilage of perishable canned cured meats by psychrotrophic clostridia seems to be more common in those products that have relatively lower salt content (e.g., $\leq$3.0%) and have been in refrigerated storage for an extended period of time (e.g., several months to a year). *C. putrefaciens* has been isolated from perishable canned cured meats that exhibited gas production and a putrid odor (Tompkin et al., 2001).

Human Enteric Pathogens in Cooked Meats

Cooking of meats to achieve an internal temperature of about 160°F (71.1°C) for at least 15–20 seconds will destroy vegetative cells of pathogens. Such destruction may also be achieved using other time and temperature combinations that are considered equivalent thermal processes. Some thermoduric bacteria and bacterial spores of both pathogenic and nonpathogenic organisms are common survivors of the cooking process. Therefore, spores of pathogenic *Bacillus* spp. and *Clostridium* spp., if initially present in meats, will most likely be found in these products following cooking at time and temperature combinations that produce a pasteurizing effect. Although thorough cooking will destroy vegetative pathogens, these organisms can contaminate cooked meats during post-cook handling actions such as chilling, skinning, slicing, and portioning.

However, as long as these products are held at temperatures below 4.4°C, only the growth of psychrotrophic pathogens such as *Listeria monocytogenes* becomes a major food safety concern during prolonged refrigerated storage. The growth rate and final populations of psychrotrophic pathogens in cooked meats will be influenced by factors such as pH, a_W, content of salt, nitrite, lactate and other antimicrobial ingredients, gaseous atmosphere, temperature, and competing microflora.

Perishable Cooked Uncured Meats

A recognized microbial hazard of cooked uncured meats is *Salmonella* that might have survived in these products because of inadequate cooking or introduced into a product as a result of cross-contamination after cooking (Bryan, 1980). The growth of pathogenic *S. aureus* in cooked uncured meats contaminated by food handlers is also a food safety hazard. In addition, spores of pathogens such as *C. perfringens*, *B. cereus*, or *B. subtilis* can germinate to produce vegetative cells during slow cooling of cooked uncured meats produced in bulk quantities. Populations of *S. aureus* or *C. perfringens* at about 10^6 per gram of cooked meats may cause foodborne illness. Historically, foodborne outbreaks involving *C. perfringens* have been linked to cooked uncured meat, poultry products, and gravy (Tompkin et al., 2001). Other microbial hazards in cooked uncured meats are *C. botulinum* and *L. monocytogenes*; however, botulism has rarely been linked to consumption of cooked uncured meats in North America (Centers for Disease Control, 1990, 1996; Tompkin, 1980). Although low numbers of *L. monocytogenes* may sporadically contaminate cooked uncured meats, there has been no reported listeriosis outbreaks associated with these products. Cooked uncured meats have been implicated in foodborne illness when food preparation, hot holding, and serving in restaurants, food service facilities and homes deviate from good food handling practices (Tompkin et al., 2001).

Perishable Cooked Cured Meats

Properly refrigerated cooked cured meats will not support growth of mesophilic pathogens; therefore, aerobic plate counts of these products are not linked to potential health hazards. Commercially packaged cooked cured meats pose a very low risk of causing staphylococcal food poisoning because *S. aureus* does not grow well anaerobically when salt and nitrite are present. *S. aureus* is a poor competitor; therefore, lactic acid bacteria, the predominant spoilage microflora in commercially cooked cured meats, are likely to inhibit growth of this pathogen. Also, cooked cured meats such as luncheon meats are usually well refrigerated and refrigeration temperatures below 44°F (6.7°C) will not permit the growth of *S. aureus*.

Frequent outbreaks of staphylococcal foodborne illness have been attributed to the consumption of contaminated ham. Such outbreaks were usually caused by contamination of the cooked ham during slicing by a food handler and subsequent holding of the sliced product at ambient temperature for several hours prior to consumption. Certain factors contribute to *S. aureus* growth and enterotoxin production in ham. Those factors include depletion in numbers of competitive microflora, temperature abuse of the ham, gaseous atmosphere, and the relatively high salt tolerance of *S. aureus*. Cooking of the ham substantially reduces numbers of competitive microflora such as the lactic acid bacteria that would have otherwise inhibited growth of *S. aureus*. Holding of the cooked ham at warm temperatures for more than 2–4 hours can facilitate multiplication of the pathogen to high numbers (depending on the initial level of contamination) and subsequent production of enterotoxin. In fact, a combination of warm temperature, oxygen, and salt in the ham strongly favors the growth of *S. aureus* over other microorganism that might contaminate the cooked product (Jay et al., 2005).

Generally, cooked cured meats are rarely implicated in other types of foodborne illness apart from staphylococcal food poisoning. In spite of the frequency with which *L. monocytogenes* may occur as a post-processing contaminant in cooked cured meats worldwide, relatively few listeriosis outbreaks implicating these meat products have been reported (McLauchin, 2006). *Salmonella* and other nonsporeforming pathogens are killed during the cooking step for commercially produced cooked cured meats. However, if these products are contaminated after cooking, salmonella can survive and under favorable temperatures for growth, the pathogen can multiply and reach infectious levels in these products.

With respect to *C. botulinum*, the main factors that contribute to control of this deadly pathogen in cooked cured meats are the presence of nitrite and salt, the growth of lactic acid bacteria, and refrigerated storage conditions. Similar factors most likely apply to *C. perfringens* in cooked cured meats; however, more research involving cured meats is needed to confirm the commonly held opinion (Tompkin et al., 2001).

Shelf-Stable Canned Uncured Meat Products

The major pathogen of concern in shelf-stable canned uncured meats is *C. botulinum*. Commercially produced beef stew was implicated in a botulism outbreak (Tompkin, 1980). While commercial processing of shelf-stable canned uncured meats has been rarely associated with botulism outbreaks, several outbreaks have been linked to home canned or home bottled meats (Tompkin, 1980). Commercial thermal treatments used for preventing spoilage of these meat products are typically more severe than the 12-D canning process (e.g., a 12-log reduction in *C. botulinum* spores). Spores of *B. cereus* and *C. perfringens* are substantially less heat resistant than those of *C. botulinum*; therefore, an adequate 12-D canning process for *C. botulinum* will completely destroy spores of these pathogens. Another food safety concern is postprocessing contamination of these products with pathogens that may enter defective containers during cooling. Outbreaks of typhoid fever in the United Kingdom from ingestion of imported canned meat products have been linked to use of nonpotable water for cooling these products (ICMSF, 2000). As a postprocessing contaminant, *Salmonella* is therefore recognized as a hazard in shelf-stable canned uncured meats.

Shelf-Stable Canned Cured Meat Products

As previously stated for shelf-stable canned uncured meats, a pathogen of concern in shelf-stable canned cured meat products is *C. botulinum*. For this reason thermal treatments and ingredient composition of shelf-stable canned cured meat products are designed to prevent growth of this deadly pathogen. The stability of these products depends on factors such as level of nitrite and salt, low numbers of indigenous bacterial spores in the raw materials, and heat treatment that inactivates many of the spores (Tompkin, 2005). The presence of adequate amounts of nitrite and salt in these products will inhibit germination and outgrowth of spores that survive the heat treatment. Postprocessing contamination with other pathogens such as *Salmonella* can occur during cooling of these products with nonpotable cooling water. In fact, cans of corned beef that were cooled in river water were implicated in outbreaks of typhoid fever (Meers and Goode, 1965; Milne, 1967).

Perishable Canned Cured Meat Products

Perishable canned cured meats typically contain low numbers ($\leq 10^2$ CFU/g) of viable mesophilic aerobic and anaerobic sporeformer survivors that do not germinate and grow at refrigeration temperatures. The presence of unusually high numbers ($\geq 10^3$ CFU/g) of mesophilic anaerobic sporeformers that will not grow at $\leq 50°F$ ($\leq 10°C$) is indicative of temperature abuse of these meat products. In this regard, a potential botulism hazard must be taken into consideration (Tompkin et al., 2001). Upon opening a suspect can, the detection of a putrid odor may be an initial indication of a botulism risk; however, the analyst needs to be familiar with putrid odor produced by putrefactive anaerobes (Tompkin et al., 2001).

References

Ball, C.O., and Olson, F.C.W. 1957. *Sterilization in Food Technology: Theory, Practice and Calculations*. McGraw-Hill, New York.

Bell, R.G., and De Lacy, K.M. 1982. The role of oxygen in the microbial spoilage of luncheon meat cooked in a plastic casing. *Journal of Applied Bacteriology* 53:407–411.

Bell, R.G., and De Lacy, K.M. 1983. A note on the microbial spoilage of undercooked chub-packed luncheon meat. *Journal of Applied Bacteriology* 54:131–134.

Blickstad, E., and Molin, G. 1983. The microbial flora of smoked pork loin and frankfurter sausage stored in different gas atmospheres at 4°C. *Journal of Applied Bacteriology* 54:45–56.

Broda, D.M., DeLacy, K.M., and Bell, R.G. 1998a. Efficacy of heat and ethanol spore treatments for isolation of psychrotrophic *Clostridium* spp. associated with spoilage of vacuum-packed chilled meats. *International Journal of Food Microbiology* 39:61–68.

Broda, D.M., DeLacy, K.M., and Bell, R.G. 1998b. Influence of culture media on the recovery of psychrotrophic *Clostridium* spp. associated with spoilage of vacuum-packed chilled meats. *International Journal of Food Microbiology* 39:69–78.

Bryan, F.L. 1980. Foodborne diseases in United States associated with meat and poultry. *Journal of Food Protection* 43:140–150.

Centers for Disease Control. 1990. Foodborne disease outbreaks – Five year summary, 1983–1987. In: CDC Surveillance Summaries. *Morbidity and Mortality Weekly Report* 39(SS-1):15–57.

Centers for Disease Control. 1996. Surveillance for foodborne disease outbreaks – United States, 1988–1987. In: CDC Surveillance Summaries. *Morbidity and Mortality Weekly Report* 45(SS-5):1–66.

Holley, R.A., Guan, T.Y., Pierson, M., and Yost, C.K. 2002. *Carnobacterium viridans* sp. nov., an alkaliphilic, facultative anaerobe isolated from refrigerated, vacuum packed bologna sausage. *International Journal of Systematic and Evolutionary Microbiology* 52:1881–1885.

ICMSF (International Commission on Microbiological Specifications for Foods). 2000. Meat and meat products. In: *Microorganisms in Foods 6,*

Microbial Ecology of Food Commodities. Aspen Publishers, Gaithersburg, MD, pp. 1–74.

Ingram, M., and Roberts, T.A. 1976. The microbiology of the red meat carcass and the slaughterhouse. *Royal Society Health Journal* 96:270–285.

Jackson, T.C., Marshall, D.L., Acuff, G.R., and Dickson, J.S. 2001. Meat, poultry, and seafood. In: Doyle, M.P., Beuchat, L.R., and Montville, T.J. (eds), *Food Microbiology, Fundamentals and Frontiers*, 2nd edn. ASM Press, Washington, DC, pp. 91–109.

Jay, J.M., Loessner, M.J., and Golden, D.A. 2005. *Modern Food Microbiology*, 7th edn. Springer Science and Business Media, New York.

Kalchayanand, N., Ray, B., and Field, R.A. 1993. Characteristics of psychrotrophic *Clostridium laramie* causing spoilage of vacuum-packaged refrigerated fresh and roasted beef. *Journal of Food Protection* 56:13–17.

Kalinowski, R.M., and Tompkin, R.B. 1999. Psychrotrophic clostridia causing spoilage in cooked meat and poultry products. *Journal of Food Protection* 62:766–772.

Kempton, A.G., and Bobier, S.R. 1970. Bacterial growth in refrigerated, vacuum-packed luncheon meats. *Canadian Journal of Microbiology* 16:287–297.

Makela, P., Schillinger, U., Korkeala, H., and Holzapel, W.H. 1992. Classification of ropy slime-producing lactic acid bacteria on DNA–DNA homology, and identification of *Lactobacillus sake* and *Leuconostoc amelibiosum* as dominant spoilage organisms in meat products. *International Journal of Food Microbiology* 16:167–172.

McLauchin, J. 2006. Listeria. In: Mortarjemi, Y., and Adams, M. (eds), *Emerging Foodborne Pathogens*. Woodhead Publishing Ltd., Cambridge, England, pp. 406–428.

Meers, P.D., and Goode, D. 1965. *Salmonella typhi* in corned beef. *Lancet* i:426.

Milne, D. 1967. The Aberdeen typhoid outbreak (1964) Report of the Departmental Committee of Enquiry, Scottish Home and Health Department, Edinburgh.

National Canners Association. 1968. Process calculations. In: *Laboratory Manual for Food Canners and Processors*. National Canners Association. AVI Publishing, Westport, CT, p. 220.

Niven, C.F. 1952. Significance of the lactic acid bacteria in the meat industry. In: *Proceedings of the American Meat Institute's Fourth Research Conference*. AMI, Chicago, IL.

Pearson, A.M., and Gillett, T.A. 1999. *Processed Meats*. Kluwer Academic Publishers, New York.

Ray, B. 2001. *Fundamental Food Microbiology*, 2nd edn. CRC Press, New York.

Roberts, T.A., Macfie, H.J.H., and Hudson, W.R. 1980. The effect of incubation temperature and site of sampling and assessment of the numbers of bacteria on red meat carcasses at commercial abattoirs. *Journal of Hygiene* 85:371–380.

Tompkin, R.B. 1980. Botulism from meat and poultry products – A historical perspective. *Food Technology* 34:229–236, 257.

Tompkin, R.B. 1986. Microbiology of ready-to-eat meat and poultry products. In: Pearson, A.M., and Dutson, T.R. (eds), *Advances in Meat Research*, Vol. 2. AVI Publishing, Westport, CT, pp. 89–121.

Tompkin, R.B. 1995. The use of HACCP for producing and distributing processed meat and poultry products. In: Pearson, A.M. and Dutson, T.R. (eds), *HACCP in Meat, Poultry and Fish Processing, Advances in Meat Research Series*, Vol. 10. Blackie Academic & Professional, New York, pp. 72–108.

Tompkin, R.B. 2005. Nitrite. In: Davidson, P.M., Sofos, J.N., and Branen, A.L. (eds), *Antimicrobials in Food*, 3rd edn. CRC Press, Taylor & Francis Group, Boca Raton, FL, pp. 169–236.

Tompkin, R.B., McNamara, A.M., and Acuff, G.R. 2001. Meat and poultry products. In: Downes, F.P., and Ito, K. (eds), *Compendium of Methods for Microbiological Examination of Foods*, 4th edn. American Public Health Association, Washington, DC, pp. 463–471.

Whiteley, A.M., and D'Souza, M.D. 1989. A yellow discoloration of cooked cured meat products: Isolation and characterization of the causative organism. *Journal of Food Protection* 52:392–395.

Yang, R., and Ray, B. 1994. Prevalence and biological control of bacteriocin-producing psychrotrophic leuconstocs associated with spoilage of vacuum-packaged meats. *Journal of Food Protection* 57:209–215.

CHAPTER 3

Fundamentals of Continuous Thermal Processing

Donald Burge, Gold'nPlump Poultry

Continuous Versus Batch Processes

A single modern continuous processing line for ready-to-eat (RTE) meat products produces 1820 kg (approximately 4,000 lb) to 4540 kg (approximately 10,000 lb) of product per hour. So, an average plant with four processing lines might produce 12725 kg (approximately 28,000 lb) of product every hour. Annualized, this one plant would produce 26.5 million kilograms (58.3 million pounds) of RTE meat. Estimating the total annual consumption for the entire USA is more difficult than for single plants or categories of RTE meats due to a lack of published information. I can estimate a U.S. total based on U.S. census figures and total per capita meat consumption from the United States Department of Agriculture (USDA) Economic Research Service (ERS). According to the U.S. Census Bureau's "POPClock" (U.S. Census Bureau, 2007), the total U.S. population in July 2007 was 302461017. The latest USDA ERS' (USDA ERS, 2007) total meat consumption figures available for the USA is for the year 2005. The USDA ERS' estimate is 90.9 kg (200 lb) of red meat, poultry, and fish consumed per year per person. Therefore, if a very conservative estimate of 10% of all meat, poultry, and fish consumed in the USA is RTE; then, we consume 2.75 billion kilograms (6.0 billion pounds) of RTE meat and fish products in the USA alone. The sheer volume produced simultaneously creates a huge economic opportunity and food safety risk.

If the RTE meat products produced are small solids then the primary process utilized will most likely be continuous. Pizza toppings and breaded chicken breasts would be examples of small solids. The common element of these small products is their relatively short cooking time. For fluids or large products, a batch process will be the most likely choice. Soups and bone-in ham would be examples of fluid and large solid products. The common element of large products is a relatively long cooking cycle, which makes it prohibitive to make continuous systems large enough to accommodate the cooking cycle. This limitation is both a physical and an economic barrier. Floor space in the processing plant creates the economic barrier to large thermal processing equipment. The focus of this book is RTE meat products. The focus of this chapter is RTE meat products processed on conveyorized continuous thermal processing or cooking equipment. In academia, adding thermal energy to food in order to create a food safe and more desirable product to eat is called thermal processing. Most industry personnel will use the term *cooking* in place of thermal processing. In fact, many industry personnel have "cookers" rather than ovens, steamers, etc. Note that throughout this chapter, I use the terms *cooking* and *thermal processing* interchangeably.

Cooking products on conveyorized continuous ovens possess some significant engineering challenges. Preventing leakage or exchange of gas between the oven and atmosphere is a big challenge when a conveyor must be fed through the length of the oven. So, by necessity, the oven will have openings at two ends, making it difficult to prevent exchange between the cooking environment within the oven and the outside air. The issue of maintaining the cooking environment in a continuous oven is often referred to by manufacturers as containment. Oven manufactures spend considerable resources in developing engineering solutions to deal with oven gas containment issues in continuous systems. The result has been continuous improvement in the ability of ovens and steamers to maintain the set conditions within. As containment has improved, so has the variety of possible cooking environments. The variety of cooking environments may be a blessing for those capable of using them effectively. For those who cannot, the variety of cooking conditions may be a curse. That is, there may be "too much choice."

Heating Mediums

Continuous cooking systems work using one of a number of possible fluid heating mediums. The three most common fluids are oil, water, and gas (physicists consider gas a very low viscosity fluid). When oil is used, the process is broadly termed frying. Fortunately, there are numerous books and scientific articles available covering the frying process. Most of the frying publications deal with foods other than RTE meat, but most of the basic principles of frying other foods will apply to coated meat and poultry (Bouchon et al., 2003).

Cooking with water is a relatively energy-efficient process, although the process is primarily applied to packaged meat products. Water is generally the medium used for postprocess pasteurization of packaged RTE meats. Water cooking does not allow the product to develop any color, also referred to as browning. The lack of browning limits the application of water as a cooking medium.

Atmospheric gas is the most common cooking medium for RTE meats. The gas medium utilized in cooking varies widely because the gas can nearly be devoid of water vapor to nearly pure water vapor (steam). The variation in moisture content provides for a great deal of variation in the attributes of the finished RTE meat products that may be produced. My discussion that follows will primarily focus on gas cooking media, since there is relatively less published material available for gas even though it is the most common medium.

Continuous Thermal Processing Equipment

Continuous conveyorized thermal processing equipment comes in two basic types: steamers and ovens. The distinction between the two is both heat source and the temperature range of operation. Live steam is the only heat source in a steamer. Therefore, at atmospheric pressure steamers always operate at temperatures below 100°C (212°F). Ovens may utilize steam and heated air as a heat source, allowing variation in the amount of moisture available within the oven box and heating the gas medium above 100°C.

Ovens and steamers can be built in two configurations: linear and spiral. Linear configurations accommodate small items that can be cooked quickly, generally in less than 12 minutes of total cooking time. A typical linear oven is shown in Fig. 3.1. Linear ovens have the capability for

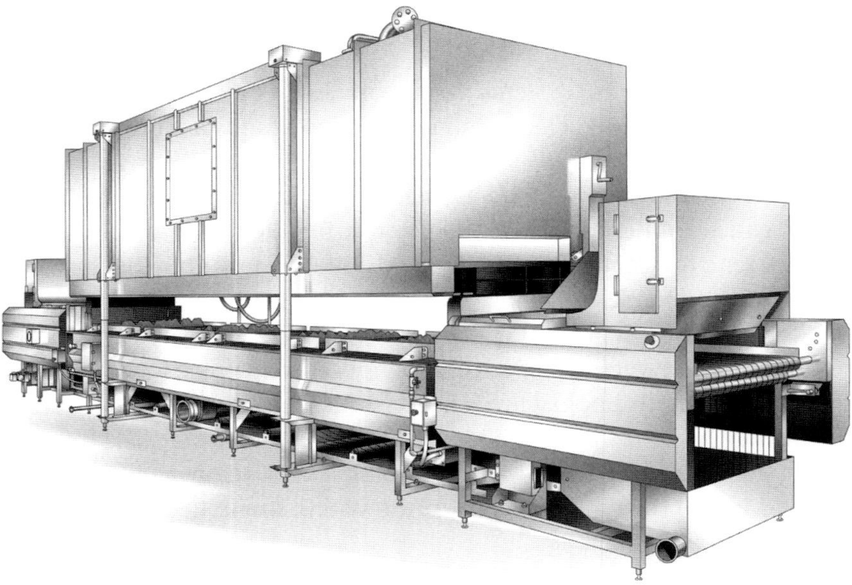

Figure 3.1. Typical linear continuous conveyorized oven. FMC FoodTech JSO IV with permission.

relatively high-velocity airflow. Typically, airflows range from 610 m/min (2,000 ft/min) to 2440 m/min (8,000 ft/min) in linear ovens. At these airflows, the ovens are generally referred to as impingement ovens. The cooking performance of linear impingement ovens is primarily a function of air temperature, air velocity, and the oven's containment capability. Containment is the ability of the oven to resist exchange between the atmosphere inside the oven box and the air outside the oven box. The better the containment the higher the amount of moisture can be maintained in the oven box.

Spiral ovens are so named because the belts appear like a spiral staircase, moving the product up or down through the oven box. Air velocities typically range from 150 m/min (500 ft/min) to 225 m/min (750 ft/min). The primary advantage of the spiral design is the ability to economically cook at longer times utilizing less floor space. Of course, any type of oven can be set up in series with any other type creating processing lines capable of producing a myriad of thermal process systems. In addition, some hybrid designs with elements of linear and spiral designs combined are available, as shown in Fig. 3.2.

Figure 3.2. Typical spiral continuous conveyorized oven. FMC FoodTech GCO II with permission.

Continuous conveyorized microwave ovens generally perform well with a specific and relatively small segment of RTE meats. The largest application is for thermal processing bacon.

Oven Variables

No matter where in the world continuous ovens operate, the owner's primary concerns are food safety, consistency, and economics of the process. Food safety is a prerequisite to doing business, and the fundamentals are well covered in published literature as well as in other chapters of this book. The focus of the remainder of this chapter will be the process economics, process consistency, and quality of products produced from continuous thermal processes. Most operations personnel are driven by economics. The impact of small changes in process efficiency can have a large economical impact. A typical modern continuous line will process 1,820–4,540 kg/h. At a value of $4.40 per kg, a 1% change in the cooking yield will range in value from approximately $80.00 to $200.00 per hour. These costs add up quickly and represent approximately $166,000.00–$416,000.00 on an annualized basis for a single processing line. These economics drive great efforts to "optimize" processing lines. In order to truly optimize any processing line, the variables the operator can set, or independent, variables must first be defined. I mentioned economics as the output, or

Table 3.1. Oven controls available on a typical linear and spiral oven

Control	*Linear Oven*	*Spiral Oven*
Temperature	Yes	Yes
Dwell time	Yes	Yes
Moisture content	Yes	Yes
Air velocity	Yes	Yes
Finger height	Yes	No
Air balance	Yes	No
Air direction	Yes	Yes
Nonpowered vent	Yes	No
Powered damper	Yes	Yes
Gate position	Yes	Yes

The descriptions of the controls above will vary by manufacturer.

dependent, variable of interest previously. However, other dependent variables such as product taste, appearance, and overall quality must also be considered. Table 3.1 contains a compilation of the most common settings available to an operator of a typical linear and spiral oven.

The large number of options available in a typical oven as shown in Table 3.1 leads to a problem. The problem is how can you possibly optimize any system given the number of choices and options available? To make this overabundance of choice problem even worse, the independent variables manipulated by the controls listed in Table 3.1 often interact. The implication of the interaction of the control variables available is that there will not be a single optimal setting for any given product. I will discuss how to deal with interacting variables in the experimental section that follows. Overcoming variable interaction is probably the single greatest challenge in process optimization of oven systems.

Fortunately, experimental methodology deals effectively with any manufacturing system, even those with multiple independent variable inputs. Factorial experiments can be designed to include multiple input, independent, variables simultaneously. It is also possible to include multiple output, dependent, variables in the design. The ultimate result of properly designed and executed oven experiments can be a truly optimized system. The information obtained with factorial (and other designed) experiments will define whether the independent variables are actually independent, or whether they interact. The distinction is critical. If the independent variables do not interact there will be a single best, or optimal, setting for that variable. If the independent variables interact there will not be a single optimal setting for any interacting variable. Understanding and dealing with this statistical fact is absolutely the key to optimizing any oven system, as is the case for almost any industrial manufacturing process.

Thankfully, all the settings listed in Table 3.1 can be condensed into four basic variables or factors. These factors are temperature, humidity, time, and air velocity. By condensing the control variables to four fundamental independent variables it is feasible to design practical experiments with these four factors.

Experimental Methodology

To design, execute, and analyze any experiment in an industrial setting on a continuous process is a considerable challenge. The only truly effective methodology to deal with multifactor systems is design of experiments (DOEs). The DOE body of knowledge is well covered by Juran and Godfrey (1999) and Moen et al. (1999). To illustrate the power and usefulness of DOE, we will take the multiple independent variables from a linear oven in Table 3.1 and design some example experiments as were previously presented by Burge and Gunawardena (1997). In any process optimization where there are more independent variables than can be designed into a single experiment, some method to reduce the independent variables must be employed. Screening experiments are commonly employed to reduce the number of variables in an experiment. Juran and Godfrey (1999) cover many available options. In the example, I present in this chapter, Burge and Gunawardena (1997) chose to group the oven control options by the fundamental effect they have on the product. So, both fan speed and impingement nozzle height change the gas velocity on the product surface.

As I stated earlier, all the control options available manipulate either the temperature, moisture level, fluid velocity, or dwell time. The measurement of each of these options is straightforward, with the exception of moisture level. Oven manufacturers differ on the best method to quantify the amount of moisture inside an oven. The two most common measures are relative humidity (RH) and percent moisture by volume (%MV). The RH measurement is common in batch processes, like smokehouses, and can be measured using a simple wet-bulb thermometer. The %MV requires a more complicated measuring device, usually based on conductivity. The %MV is easier to interpret than that of RH inside an oven, since %MV is simply the percentage of total volume within the oven occupied by water vapor. So, if the internal oven chamber had a total volume of 300 ft^3 and contained 150 ft^3 of water vapor, the %MV would be 50.

Table 3.2. Treatment arrangement for a 3 × 3 × 3 full factorial oven experiment

		Air Velocity m/min (ft/min)		
Oven Temperature °C (°F)	*Percent MV*	*686 (2,250)*	*1,029 (3,375)*	*1,372 (4,500)*
190.6 (375)	40			
190.6 (375)	60			
190.6 (375)	80			
218.3 (425)	40			
218.3 (425)	60			
218.3 (425)	80			
246.1 (475)	40			
246.1 (475)	60			
246.1 (475)	80			

If we assign temperature, moisture level, and fluid velocity as factors in a factorial experiment then an arrangement of treatments arises as shown in Table 3.2. Note that every factor is present at three levels. So, temperature, for example, has three levels: 190.6, 218.3, and 246.1°C. This table describes a 3 × 3 × 3 factorial experiment, as we have three factors with three levels each. This example is a subset of the larger 2 × 3 × 3 × 3 experiment described below. We also know, based on the arrangement in Table 3.2, that there will be a total of $3^3 = 27$ possible factor level combinations. So, when executing this experiment we would take at least two measurements at each possible combination. Therefore, if we decide to use twofold replication, we would make a total of $27 \times 2 = 54$ measurements. The analyses, I show below, are based on the experiment shown with sixfold replication: $27 \times 6 = 162$ measurements. I have used the series of experiments described below as part of the evaluation of a new linear oven design.

We ran one experiment, as arranged in Table 3.2, at 3.8-minute dwell time combined with an experiment at 5.1-minute dwell time, which means the total experimental design is a 2 × 3 × 3 × 3 factorial. In the last experiment we performed the same 2 × 3 × 3 × 3 design, but we adjusted dwell time in order to insure a constant endpoint temperature 76.7°C (170°F). The response variable for the first 2 × 3 × 3 × 3 experiment was internal, or core, temperature of the product leaving the oven. This measures the amount of energy delivered to the center of the product. In the second experiment, where we cooked to a set temperature, we measured

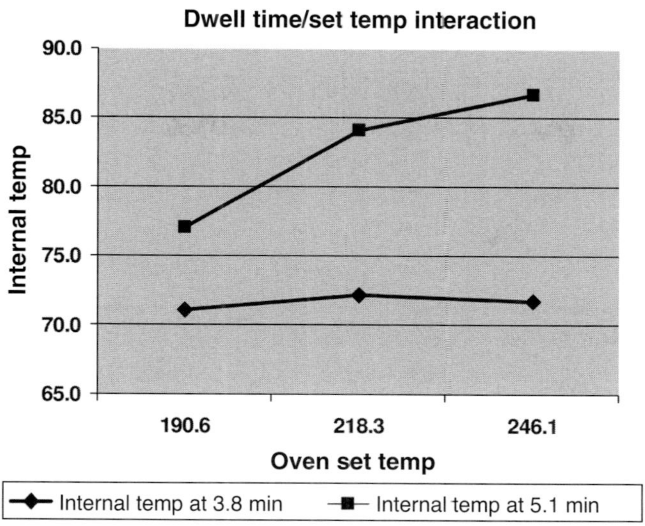

Figure 3.3. Two-way interaction between oven set temperature (°C) and oven dwell time (minutes). The internal temperature is the temperature of the product in °C. This graph is based on data from Burge and Gunawardena (1997).

the initial and final product weight in order to calculate the product yield as finished cooked weight divided by the initial raw weight × 100%. Note that in the discussion below, we break the large 2 × 3 × 3 × 3 experiment in two 3 × 3 × 3 factorials. We did this now and when presenting the original paper to aid in interpretation of the results.

The overall results of all the experiments described above are contained in Tables 3.3–3.5 and Fig. 3.3. Table 3.3 contains the mean internal temperatures of formed chicken patties processed in linear oven for 3.8 minutes. Raising the oven temperature from 190.6 to 246.1°C (100°F)

Table 3.3. Mean internal product temperature at an oven dwell time of 3.8 minutes

Oven Temperature (°C)	*Internal Temperature (°C)*	*Gas Velocity (m/min)*	*Internal Temperature (°C)*	*% MV*	*Internal Temperature (°C)*
190.6	71.1	686	70.4	40	63.6
218.3	72.2	1,029	71.0	60	72.2
246.1	71.7	1,372	73.3	80	78.9

Adapted from Burge and Gunawardena (1997).

Table 3.4. Mean internal product temperature at an oven dwell time of 5.1 minutes

Oven Temperature (°C)	*Internal Temperature (°C)*	*Gas Velocity (m/min)*	*Internal Temperature (°C)*	*% MV*	*Internal Temperature (°C)*
190.6	77.0	686	81.2	40	76.4
218.3	84.1	1,029	81.5	60	82.6
246.1	86.7	1,372	85.2	80	88.8

Adapted from Burge and Gunawardena (1997).

did not change mean internal product temperature. The difference between 71.1 and 72.2°C is not statistically different. As gas velocity increased from 686 to 1372 m/min the internal product temperature increased from 70.4 to 73.3°C, which is statistically significant though numerically small. By far the largest effect was changing percent moisture by volume. The increase in MV from 40 to 80% resulted in an increased internal temperature of 15.3°C (27.5°F). It is counterintuitive to most oven operators that increasing oven temperature 55.5°C and gas velocity 686 m/min caused little effect on internal product temperature. These results are even more surprising after looking at the results at 5.1-minute dwell time in Table 3.4.

The results of the experiment at 5.1 minutes are shown in Table 3.4. At 5.1-minute dwell time, an increase in oven temperature from 190.6 to 246.1°C becomes a statistically and numerically significant factor. The increase in oven set temperature increased the internal product temperature by 9.3°C (16.7°F). Gas velocity is still significant, especially the increase from 1029 to 1372 m/min. The increase in %MV is also significant, although the magnitude is less than at 3.8-minute dwell time. At 3.8-minute dwell time, an increase in the MV from 40 to 80% resulted in a 15.3°C (27.5°F) rise in internal product temperature, as shown in Table 3.3. The same %MV increase we used at 3.8 minutes now at 5.1 minutes resulted in a 12.4°C (22.3°F) rise in internal product temperature. Hence,%MV had less effect at a 5.1-minute dwell time than at a 3.8-minute dwell time.

Probably the most striking difference in the results in Tables 3.3 and 3.4 is the effect of oven temperature. The increase in oven set temperature increased the internal product temperature by 9.7°C at 5.1 minutes, but had no significant effect at 3.8-minute dwell time. This apparent discrepancy may, at first, seem like an error, although I found the results repeatable. In fact, this phenomenon is a variable interaction. In order to understand how the interaction works, I will recombine the results of

Tables 3.3 and 3.4. This is possible because I executed these two studies simultaneously, as previously mentioned. That is, the study was actually a 2 × 3 × 3 × 3 factorial that I broke into two separate studies for interpretation. When I recombine the two-way variable interaction between oven dwell time and oven set temperature, the interaction between dwell time and oven set temperature is graphed in Fig. 3.3. The interpretation of this interaction is as follows. At 3.8-minute dwell time, an increase in the oven set temperature from 190.6 to 246.1°C has no significant effect on the internal product temperature. This is the bottom line in Fig. 3.3. At 5.1-minute dwell time, an increase in the oven set temperature now significantly increases the internal product temperature. This is the upper line in Fig. 3.3. I will discuss the mechanism behind this result later in the chapter.

For now, the importance of this result is that the interaction limits the global nature of the conclusions we can draw from the experimental results. It is incorrect to conclude that increasing oven set temperature will always increase the internal temperature of the product running through the oven. The result of the adjusting oven dwell time is said to be dependent on the dwell time. Hence, you must know the dwell time in order to predict the effect of adjusting oven temperature. If these results were independent then adjusting the oven set temperature would have a similar effect regardless of dwell time.

I have observed at first hand the practical implication of oven variable interactions in many plants that process RTE meat. Many times the operators and/or line leads from different shifts disagree over the effect of changing a control variable on their oven. Sometimes they also disagree on the optimal setting. These disagreements may baffle the plant operations group when the managers or operators from both shifts have years of experience and demonstrate competence in operating the lines. It may well be the case when variables interact that both shifts are correct. This is possible because the optimal setting of the variable in question (oven temperature in the example above) may depend on the setting of one or more other variables available to the operators. In these cases an understanding of the underlying science behind the process can help explain why the interactions occur. In the case of thermal processing of RTE meat the underlying science is heat and mass transfer, which are also covered in detail elsewhere in this book. Understanding the heat and mass transfer events during cooking can explain the basis for the interaction.

The oven control variables temperature, gas velocity, and percent moisture by volume also demonstrate extensive economic effects. Table 3.5 shows how these variables impact on the cooked yield of a formed chicken

Table 3.5. Mean cooked yield percentage after reaching a mean internal temperature of 76.7°C

Oven Temperature (°C)	*Cooked Yield (%)*	*Gas Velocity (m/min)*	*Cooked Yield (%)*	*% MV*	*Cooked Yield (%)*
190.6	91.6	686	92.4	40	86.7
218.3	91.0	1, 029	90.6	60	91.6
246.1	90.2	1, 372	89.8	80	94.4

Adapted from Burge and Gunawardena (1997).

pattie. The variables can be ranked from greatest impact on yield to least impact on yield as follows: %MV > gas velocity > oven set temperature. Percent moisture by volume increased oven cooked-yield by 7.7% when increased from 40 to 80%. To fully appreciate the impact of this change, keep in mind the economic impact mentioned earlier in this chapter. That is, a 1% change in oven yield on a modern continuous process line may affect return of $166,000.00–$416,000.00 on an annualized basis for a single processing line.

Heat and Mass Transfer and Thermal Processing

The experimental evidence outlined earlier combined with other published and unpublished results has led to a theory of changes that occur as the product moves through an oven during the thermal process. This theory was explained by Burge and Gunawardena (1997). We predicted that a product traversing an oven will go through three phases or zones as illustrated in Fig. 3.4. The three zones are condensational mass transfer dominance, transition, and convection heat transfer dominance. When the cold product first enters the oven it is cold relative to the impingement gas. The result is moisture condensation on the product surface. This results in rapid heat and mass transfer to the product. This phase of cooking will continue until the surface temperature of the product reaches the dew point of the impingement gases. The length of time (which is also distance) product spends in the first phase is dependent on the initial product temperature, product properties, and the moisture level in the impingement gas. Once the product surface temperature reaches the dew point in the oven, condensation stops. It is likely that the product will

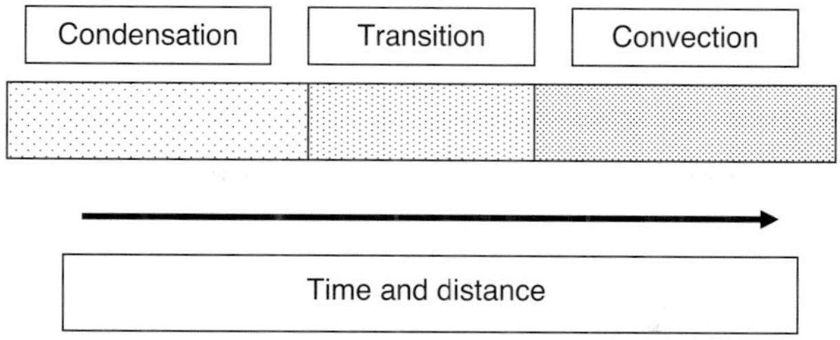

Figure 3.4. Graphic representation of the three heat and mass transfer zones meat may pass through during cooking.

have a good deal of surface moisture depending on the product from the condensation portion of the cooking process. With the surface above the dew point moisture will begin to evaporate rapidly driven by the hot impingement gases. This cools the surface back below the dew point, and condensation will reoccur. The product may cycle back and forth between condensation and evaporation on the surface in this transition zone. This cycling back and forth between heat and mass transfer has been reported in the scientific literature. Millsap and Marks (2005) noted the condensing to convection transition caused instability in their calculated heat transfer coefficients. Once the surface remains above the dew point of the oven gases the predominant cooking mechanism is convective heat transfer. In this mode the gas temperature and velocity become particularly important. All the surface browning achieved in the process will occur during convective dominant heat transfer. The rate of browning will be driven by both surface drying and surface temperature. Surface drying concentrates the reactants in the browning reaction. Increased surface temperature provides additional energy driving the reaction rate up as surface temperature increases.

Using this model it is possible to explain the interaction of dwell time and oven set temperature shown in Fig. 3.4. In the first portion of the cooking process condensational heat and mass transfer predominate. Millsap and Marks (2005) argue that this phase is solely a mass transfer phenomenon, and should be termed condensational mass transfer. This process is basically independent of temperature in the range studied 191–246°C (375–475°F). Hence, at 3.8-minute dwell time, the internal temperature is independent of oven temperature; and at 5.1-minute dwell time, the product has gone far enough to get to the stage where convectional heat transfer predominates. Convection is dependent on both oven temperature and gas velocity. Hence, the internal product temperature increases as oven temperature rises.

Product Quality Considerations

For the processor who masters the experimental methodology and an understanding of heat transfer, obtaining the desired meat product becomes a matter of routine. The processing conditions the processor chooses affect product quality on three levels. These are product consistency, product appearance and taste, and finished product composition.

Product consistency is a basic premise of successful businesses in many sectors and around the world. Consumers enter quick-service restaurants on the East Coast and the West Coast expecting the same eating experience day in and day out. The processor must "control" their process in order to deliver the consistent quality expected by their customers and the ultimate consumer.

As a general rule, continuous thermal processes that rely on higher moisture impingement gases will deliver internal temperatures and products that are more consistent. I define consistency here on a statistical basis. In other words, a lower standard deviation equates to a lower variance and more temperature and product composition consistency. Even with the oven process conditions set to reduce variability, the entire process must be controlled. For example, if a formed product is being cooked on a continuous line, I have observed that a small change of 1.1°C (2.0°F) in raw temperature at the front of the line can result in a large 9.4°C (17°F) temperature change after thermal processing. So, a small temperature shift on the raw side of the process can suddenly result in an undercooked product at the end of the processing line. The consequences of having undercooked product are so severe that operators may severely overcook product to compensate for a lack of consistency in the process.

The best methodology to prevent the overcompensation of operators to insure that product is fully cooked is to employ statistical process control (SPC). With SPC even small shifts anywhere in the entire thermal processing line can be detected, and then prevented or compensated for. The use of SPC also discourages operators from making process adjustments that are not necessary. This prevention of "tampering" as it is referred to in process control will result in a much smoother and less variable processing line operation. Most oven operators feel the need to adjust process conditions, because they get paid to do so. It is not uncommon for different shifts to have significantly different equipment setting to produce the same end product. Frequently, this results in greatly increased variability in the total process. Both Juran and Godfrey (1999) and Moen et al. (1999) provide an excellent "how to" on the use of SPC in industrial process control.

The most obvious effect of cooking on product quality is the change in taste, texture, and appearance. Hopefully, few of us have tasted raw meat and poultry to compare it directly with RTE products, but the changes that occur during cooking produce what we generally consider desirable results. The gross chemical and physical changes are covered well in *Food Chemistry* textbooks such as Fennema (1996). Foegeding et al. (1996) explained the chemical basis of flavor, texture, appearance, and chemical compositional changes that occur during the cooking of meat products. Meat flavor changes during cooking due primarily to the breakdown of fats and amino acids. According to Foegeding et al. (1996), the compounds resulting from the addition of thermal energy during cooking include aldehydes, keytones, alcohols, sulfides, and mercaptans. All these compounds potentially change the flavor of raw versus the flavor we know in RTE meat. The denaturation of structural proteins such as actin and myosin leads to the stiffening in texture of RTE meat. As meat and poultry proteins denature they also become less translucent and whiter in appearance. In addition, the denaturation and subsequent oxidation of myoglobin leads to the characteristic change in appearance during cooking. Cooking adds a large amount of thermal energy to the product. This added thermal energy drives a wide array of chemical changes during cooking. I have mentioned protein denaturation, but some other significant changes also occur. These include the inactivation of enzymes, color reactions (e.g., Maillard browning), redistribution of fat, the breakdown of collagen to gelatin, and a general reduction in the ability of the muscle to hold moisture.

The flavor, texture, and changes in appearance that occur during cooking are obvious once the product is consumed. Changes in composition are not obvious and have been the subject of scientific research. Proctor and Cunningham (1983) reported on compositional changes in coated and uncoated chicken breast and thighs during cooking. The chicken breast or thighs were cooked by baking, broiling, microwaving, pan-frying, or low-pressure deep-fat frying. No commercial equipment was utilized in this study, but the results are still quite interesting to processors. Proctor and Cunningham (1983) ranked the cooking methods with regard to total cooking loss. For uncoated breast, the methods range from 23.5 to 45.6% loss depending on the cooking method. The rankings of the methods from least to most loss were baked, broiled, microwaved, pan-fried, with deep-fat fried showing the greatest loss. The results were different for uncoated thighs. The total loss for thighs ranged from 29.5 to 52.1%. The rankings of the cooking methods from least to most loss were broiled, baked, pan-fried, deep-fat fried, with microwaving having the highest loss.

Proctor and Cunningham (1983) explained the difference in product composition by cooking method based on the surface porosity of the product. They stated that frying resulted in a more porous surface allowing moisture to escape more easily. They contend that baking and broiling cause the surface to dry and become impermeable to water. This surface porosity theory may or may not fully account for moisture that escapes from the product as moisture vapor as opposed to moisture that escapes as a liquid. There were also two important components of the cooking conditions not reported in this study. There is no mention of the environmental moisture level, which home style cooking equipment does not have the capability of measuring. Hence, they were not reported. Proctor and Cunningham (1983) also did not report the internal product temperature after cooking. They further stated that the difference in cooking performance between thighs and breast was due to the higher initial moisture and fat content of the thighs. They did not mention differences (if there were any) in the melting profiles of breast versus thigh fat.

The variable cook loss that Proctor and Cunningham (1983) observed, resulted in variable composition of the finished cooked chicken. For the uncoated breast and thighs, broiling produced the highest moisture content at 69.08%. Microwaving resulted in the highest protein content at 30.40%. Not surprisingly, the highest crude fat of 6.82% occurred in breast and thighs that were deep-fat fried. The proximate composition of the products in this study produced after cooking showed a great deal of variation. This is partly due to the inclusion of frying methods along with dry heat cooking methods. Commercial thermal processes, with the exclusion of frying, will not add any material to the product during the thermal process. Hence, all the quality and compositional difference result from the differential loss of water, protein, fat, and ash during the process.

Conclusion

For many processors of RTE meat products the thermal processing lines they own may seem somewhat like an unknown black box where the raw product goes in one end and cooked product comes out the other. The line operators may even seem like shamans, magically adjusting the ovens to get the desired results. Of course, this need not be the case. Through the application of careful measurement utilizing DOE, SPC, and a good understanding of heat and mass transfer it is possible to reliably control the cooking process. More importantly, the processor can determine a truly optimized process dealing with all the system variables simultaneously.

The processor will also be able to predict the changes in product appearance and organoleptic quality associated with various processing conditions. The final RTE product produced will be food that is safe, consistent, economical, and world class while meeting customer and consumer requirements.

References

Bouchon, P., Aguilera, J.M., and Pyle, D.L. 2003. Structure oil–absorption relationships during deep-fat frying. *Journal of Food Science* 68(9):2711–2716.

Burge, D.L., Jr., and Gunawardena, R. 1997. The effect of moisture level, fan speed, and oven temperature on process parameters and finished product attributes in a commercial linear oven system. *Presented 57th Annual Meeting of the Institute of Food Technologist*, Orlando, FL, June 14–18.

Fennema, O.R. (ed.) 1996. *Food Chemistry*, 3rd edn. Marcel Dekker, New York.

Foegeding, E.A., Lanier, T.C., and Hultin, H.O. 1996. Characteristics of edible muscle tissue. In: Owen, R.F. (ed.), *Food Chemistry*. Marcel Dekker, New York, pp. 879–942.

Juran, J.M., and Godfrey, A.B. (eds) 1999. *Juran's Quality Handbook*, 5th edn. McGraw-Hill, New York.

Millsap, S.C., and Marks, B.P. 2005. Condensing–convective boundary conditions in moist air impingement ovens. *Journal of Food Engineering* 70:101–108.

Moen, R.D., Nolan, T.W., and Provost, L.P. 1999. *Quality Improvement Through Planned Experimentation*, 2nd edn. McGraw-Hill, New York.

Proctor, V.A., and Cunningham, F.E. 1983. Composition of broiler meat as influenced by cooking methods and coating. *Journal of Food Science* 48:1696–1699.

U.S. Census Bureau. 2007. U.S. POPClock Project. http://www.census.gov/population/www/popclockus.html.

USDA ERS. 2007. Briefing Room. http://www.ers.usda.gov/Breifing/Consumption.

CHAPTER 4

Thermal Processing of Slurries

Darrell Horn and Daniel Voit, Blentech Corporation

Challenges of Heating Slurries

Over the past decades, there has been a significant surge of interest in the thermal processing and cooking of products in the food industry. This trend has been particularly strong in the meat industry where, in previous years, fresh and frozen commodities played the major role in the marketplace. An increasing number of meat processors, who traditionally packed their products with minimal processing for sale, have begun to guide their business in the direction of cooking more elaborate and exotic products such as ethnic-inspired ready meals. Thermal processing allows these processors to move in a number of directions to add valve and enhance safety. This chapter explores some of these directions.

Kitchen wisdom has taught us that cooking can be used as a means of adding value and extending use of meat products, which are by nature an expensive food ingredient. For the processor, profits for the basic fresh and frozen meat cuts are limited because there are many highly efficient competitors. Whereas cooking products, which use meats as a major or minor component, allow the processor to earn greater profits from their central product, meats. As a value-added processing technique, thermal processing allows the processor to creatively differentiate their product, separating their product offering from others in the market and can help to create niche markets.

Mixing the cooked meat product with added ingredients such as flavors, spices, pastas, vegetables, and other ingredients opens many different possibilities to the processor. These added ingredients often have a lower cost per pound than the meat, which reduces the processor's total cost per pound and still can increase the value per pound of the product by offering the consumer new experiences or convenience. Periodic reformulation then allows a processor to stay on top of dynamic consumer trends.

By varying the processing conditions and technologies employed, cooking can tenderize meats with high concentrations of connective tissue, sear softer cuts to naturally create flavor-rich Maillard browning compounds or simply pasteurize the product for reduction of spoilage microorganisms. As a preservation process, thermal processing can be used to achieve varying levels of safety and stability. Nearly all levels of thermal processing do result in at least some destruction of spoilage or pathogenic foodborne microorganism. This is a benefit that is desirable for both the processor and the consumer, but a nutrient loss is inevitable. Simple blanching processes can be used to inactivate enzyme activity in vegetable components and minimize spoilage microorganisms on the surface of particulate foods or post-packaged meats. These processes have been proven to extend the refrigerated shelf life of already-packaged meats, provided cooking and cooling is precisely controlled. If not properly cooled, thermal abuse can have the reverse effect and promote the growth of spoilage and pathogenic organisms. On the other hand, use of a classic retort process can render meat products commercially sterile with shelf lives exceeding 2 years.

For all of these reasons processors, within the USA and abroad, have consistently increased their use of thermal processing for meat products. Those who have moved into cooking have found that cooking different products poses many difficult challenges and different food safety considerations. Cooking can require new technical knowledge than what the processor developed in the business of processing uncooked products. An understanding of heat transfer, cooling techniques, and basic microbiology are required to understand the industrial processes.

Although thermal processing can be used on various forms of meat products, this chapter places emphasis on the cooking and cooling of food slurries. Food slurries are a broad product category of flowable products such as soups, sauces, and stews. They are a common component in most ready meal products, fresh and frozen.

Some products that become a thick fluid product when heated can also be considered slurries for the purposes of industrial processing. Ground beef, for example, is a solid when cold but as the meat is cooked its consistency changes. It becomes a fluid as the meat is heated up between 30°C (90°F) and 50°C (130°F), the myofibrillar proteins begin to denature which leads to the release of water, often referred to as a reduction in water-holding capacity (Warriss, 2000). The purged water mixes with the separating protein as it continues to denature above 50°C (130°F). As the meat products are heated past 140°F, the collagen shrinks and fat renders. When the slurry reaches 165°F, a flowable slurry is usually achieved. For the processor, this creates a pumpable slurry which simplifies product handling and is a core reason these products have been commercialized so successfully.

As another example, fruit compote, jams and jelly often used as a meat accompaniment are liquid in the cooking phase and then solidify when cooled. A vast range of examples of food slurries many containing meat can be found on most aisles in the market and include products such as stews, baby foods, and sauces which have solid particulates in a sauce matrix.

Food Safety Considerations

Reduction of spoilage or pathogenic microorganisms is always the primary objective of cooking. Successful thermal destruction of bacteria requires bringing all parts of the product up to a specific temperature for a specific time period. Thermal process conditions range from blanching and pasteurization to cooking for commercial sterility. The choice of the thermal process employed by the processor depends on their supply chain options in conjunction with their target food product. Processors who rely on ambient warehouse storage may have a great expense to develop a cold storage distribution network and thus focus on shelf stable products. Conversely, the entry capital cost for the production equipment for shelf-stable canned products is significant for a small processor of refrigerated foods.

The selection of a thermal process depends also on salability of the finished product. For example, consumers widely accept canned soups, which contain meat as a high value component. Canned products are shelf stable and with modern equipment can be rapidly produced. However, some producers seek to differentiate themselves by servicing a fresh or frozen market. This is in part due to a noteworthy decline in quality occurs in the long, high-temperature retort process. These types of products are sold under brands built on a premium, flavor texture or nutritional image and choose refrigerated or frozen storage to re-enforce this view. But such products are generally of higher price for the consumer in part because a percent of product remains unsold after the end of its shelf life.

A simple, common process, cooking, is a thermal process that ensures product is adequately heated to eliminate pathogens. The USDA establishes minimum cooking temperatures for meat products to eliminate pathogens. Whereas for the processor, it is important to understand that thermal death kinetics vary by pH, composition, formulation, type of organism, and cooking temperatures are clearly defined. For most meat products, temperatures should reach at least 71°C (160°F) at the coldest point. But the true measure of success should be validated with microbial testing to ensure zero colony-forming units (CFUs) pathogenic and coliform bacteria survive. Guidelines for total CFU should also be established for quality purposes. If products are intended for shelf-stable storage, special thermal processing techniques, packaging certified operators, and an approved process must be employed to ensure the safety of the consumers.

A blanching process can be rarely used for the meat components in a food product because this would not guarantee a completely cooked and safe meat. However, for processors of stews and soups, the process should be understood as it applies to vegetables. Blanching is largely used to preserve color and flavor quality of green vegetables. Many meat products such as chili or stew contain significant vegetable components. If these ingredients are added fresh without a blanch process, it is important to understand that they can introduce large amounts of viable microorganisms. On the other hand, if the ingredients are purchased in the common IQF form, they are typically blanched prior to freezing. The blanching process varies by the size, shape, and type of the vegetable but if often targeted to hold the product to 93°C (200°F) for 60–120 seconds. The process completion is normally validated by a negative measurement for the peroxidase enzyme widely considered to be the most heat-stable enzyme in green vegetables. Microbial standards are generally made available for the processor purchasing the ingredient. Nongreen vegetables may be heated for the elimination of other enzymes. Tomatoes are heated to eliminate polygalacturonase and pectin methylesterase which can result in thin or separated tomato sauce. By contrast, some vegetables such as onions are low in quality-reducing enzymes but have large microorganism populations. For these products, a blanching process can be used to reduce the microbial population.

For products that are a mixture of all precooked components, a basic pasteurization process is often used to further reduce the population of spoilage organism and possibly set the starch or thickeners used in the formula. The cooking process, depending on the thickener used, generally heats the product to temperatures below boiling but exceeding 71°C (160°F). This cooking process does not result in a sterile product but does kill a significant portion of pathogenic and spoilage microorganisms that may be viable in the slurry. Following pasteurization, it is critical to ensure that the slurry is rapidly cooled to chilled or freezing temperatures to maintain quality and prevent the regrowth of microorganisms. Most spore-forming microorganisms can easily survive a pasteurization process and are often implicated in foodborne illness. Two examples found in products such as soups and stews include *Clostridium perfringens* and *Bacillus cereus*. These toxin-forming bacteria commonly occur in prepared foods containing meat and gravies. Growth of these organisms can be prevented following the USDA regulations for cooking and cooling. Because these organisms can grow between 5°C (40°F) and 60°C (140°F), the processor should select technology that rapidly cools the slurry to below 5°C (40°F). This can be accomplished in the cooker or in the package. There are benefits and risks to either approach. In either case, the total cooling time should be less than 4 hours. Technologies such as vacuum cooling can accelerate the cooling of thick particulate containing slurries to less than 1 hour for large batches.

Thermal processing for commercial sterility (shelf-stable processing) requires special consideration. These processes are best addressed as a separate topic due to the complexity and significant safety risks. Process temperatures are hotter and times are longer than simpler cooking or pasteurization processes. Typical conditions heat the product in a sealed container under pressure to temperatures exceeding 121°C (250°F). The implementation of this type of processing requires greater depth of understanding about microbiology and food safety. Processors who wish to produce these types of products and do not have direct experience should contact a University Extension Program to learn about programs offered for industry outreach. Production of shelf-stable retorted products requires the filing of an approved thermal process with a thermal process authority, heat penetration validation tests, and can only be prepared in the presence of personnel who have passed a Better Process Control school.

Different Methods and Reasons for Cooking

Beyond food safety concerns, the primary reason for cooking slurry products may vary. In some cases the objective of cooking is to create a specific texture to the product such as slow cooked stew to break down the collagen connective tissue in the stew meat. A second objective of cooking may be to improve the overall quality of the product by melding different flavors such as the blending of spices in ethnic dishes such as Indian products. In all cases, it is desirable to cook the product without adding off-flavors often attributed to scorching or burning some of the ingredients on the heat exchange surface.

There are a number of ways to cook products, which are discussed previously. Vegetable or pasta products can be blanched in water. Meat products can be cooked in an oven with hot air or steam. Products can be fried at very high temperatures in a wok or deep-fried in hot oil. Each of these methods of cooking has their own technical challenges. Most require specialized cooking equipment. Although direct injection of steam is a commonly employed cooking process, this section concentrates on thermal processing of slurries with indirect heat transferred into the product through the walls of the cooking vessel.

Indirect heating of food slurries such as thick soup or soup with particulates, chili with meat and beans, fruit preserves and pie fillings, meat in BBQ sauce or cream-based custard desserts all create different challenges for the processor. Each product is unique and could warrant a separate study or discussion and many of the challenges revolve around heat transfer from the heating medium into the product. Fortunately, the heat transfer mechanics are universal and can be applied to each product in generality with great success.

To the chef or operator, these challenges reveal themselves in the form of burn-on on the walls of the cooker, overcooking or scorching of some of the product, and undercooking of pockets of product. For ready meal products, as previously discussed, undercooked pockets of product pose a significant safety risk. No undercooked pockets or pieces can be permitted to contaminate the finished product. Particularly thick products and products with large pieces must be cooked to ensure that all regions including the interior of large pieces receive an adequate thermal process. Industry has not yet developed a method for measuring the internal temperature of a particulate without penetration and direct measurement. Trials and validation tests are always necessary to confirm that for a given hold time, all pieces within the slurry are cooked. In fact, it is often the case that a surface mount or probe-type temperature sensor may detect a proper pasteurization temperature at the probe but subsequent measurement with a hand-held probe confirms pockets of significantly lower temperature. Processors using this technology often compensate by using long hold times reducing product quality. When a process is scaled up, automatic PLC recipe programs can be used to aid in the production of consistent and safe products. For HACCP record keeping, use of chart recorders connected to the RTD with frequent calibration validation tests is recommended.

Processing Factors Which Affect Heat Transfer

In order to understand the root causes of cooking problems, a discussion of the factors affecting heat transfer is necessary. There are several processing factors that will influence the transfer of heat into cooked product.

Overall, heat transfer is driven by the temperature difference between the hot side of the jacket and the cool product on the inside of the kettle. Heat transfer from steam at 177°C (350°F) into the product at 38°C (100°F) will be much faster than heat transfer if the temperature difference is only a few degrees. This temperature difference between the hot side of the cooker wall and the cool product side is called the "ΔT." The higher is the ΔT, the more efficient the heat transfer. If heating with steam, the steam temperature will vary depending on the steam pressure between 100°C and 177°C (212°F and 350°F).

On the whole, if the cooker is a well-mixed vessel then the heat transfer process can be characterized by its U-factor. The U-factor is a composite measure of resistance to heat transfer across two states. Figure 4.1 illustrates that a U-factor quantifies rate of heating through the jacket, trough, and into the product in a jacketed slurry cooker.

$$q = U\,A\,\mathrm{LMTD}$$

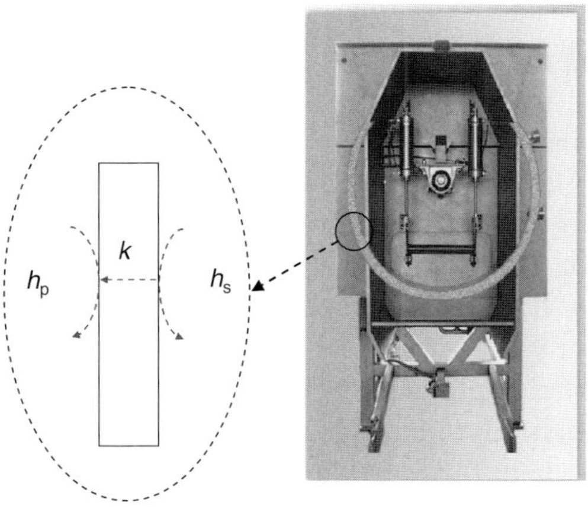

Figure 4.1. Factors affecting heat transfer (h_p, convective heat transfer coefficient jacket to product; k, thermal conductivity of cooker trough; h_s, convective heat transfer coefficient of steam to trough inner wall).

where q is the rate of heat transfer (Btu/h), A is the heated surface area (sq ft), and LMTD (F) is the log mean temperature difference.

$$\text{LMTD} = \frac{(\Delta T_{\text{max}} - \Delta T_{\text{min}})}{\text{Ln}(\Delta T_{\text{max}}/\Delta T_{\text{min}})}$$

Therefore, a U-factor can be expressed as BTU/h sq ft F and is a relative measure of rate of heating. A U-factor can be used for scale-up calculations, but is highly dependent on processing conditions so that a U-factor for one product can be very different if processed on a hemispherical kettle versus a jacketed ribbon blender. For horizontal cookers, U-factors for cooking ground meats may be as low as 50–100 BTU/h sq ft F or as high as 300 BTU/h sq ft F for thinner soups and sauces. The rate of heat transfer is highly dependent on the design of the cooker. A U-factor measured in a hemispherical kettle is often lower than a value measured in a horizontal cooker.

Because the temperature differential "ΔT" drives heat transfer, thermal oil can be used to achieve much higher rates. Thermal oil be heated much hotter than steam, and the same heat transfer factors apply. It can be heated up to 600°F using food grade heat transfer fluids. However, specialty design cookers and thermal oil heaters are needed to use this type of system. Thermal oil systems have grown in popularity in Europe and South America. In many regions, thermal oil supply is found in most ready meal plants.

Quality of Mixing: The Key to Heat Transfer

One of the most important factors in cooking is the quality of mixing. Slurries must absorb the heat from the vessel's heated jacket, but the heat energy transfer is vastly accelerated by mixing and distributing the heat from one particle in the slurry batch to the next particle with which it comes in contact. If the slurry is not mixed properly the jacket heat builds up in the product lying on the heated jacket walls, causing overcooking and eventually burned product on the cooker walls.

Most inexperienced cooks attribute burn-on to poor scraping of the cooker walls. Actually effective mixing, involving quickly and continuously moving the heated particles away from the heated jacket surfaces and distributing these heated particles throughout the batch, does more to minimize burn-on than scraping the surface. Put another way, if the heat

Figure 4.2. A dual agitator ribbon style cooking vessel with spring-mounted scrapers such as the blending cooker above can significantly improve the cooking of stiff, viscous products.

exchange surface is effectively scraped, but the heated particles are not mixed efficiently and are allowed to stay close to the heated surface of the kettle, burn-on will eventually occur anyway. On the other hand, if a cooking vessel has an effective mixing system which sweeps the heated particles of product away from the jacket surface and mixes them evenly throughout the batch, burn-on will not occur.

Since heat transfer is affected by the efficiency of the mixing of the slurry, it is important to understand the characteristics of an effective agitation system. The most effective agitation system is a horizontal shaft agitator. Figure 4.2 illustrates a range of agitators commonly used in hemispherical kettles and Fig. 4.3 illustrates agitators used in horizontal mixer cookers. For high viscosity products and products with a large portion of particulates, the vertical agitators used in hemispherical kettles often result in damage to particulates and produce weaker mixing dynamics unless higher revolutions per minute (RPM) are used.

By contrast, horizontal mixer cookers produce a more evenly mixed and cooked product with significantly less damage to particulates. The reason is that a horizontal agitator moves the product horizontally and vertically at the same time. The spokes that hold the ribbons or paddles in place impart a vertical force to the product as they rotate around the horizontal

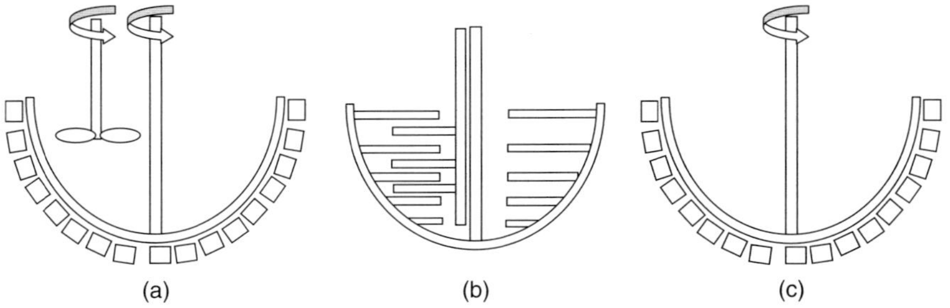

Figure 4.3. Agitators commonly used in hemispherical kettle cookers. (a) Dual motion mixer with high-speed propeller; (b) single motion mixer with static mixing/breaker bar; (c) single motion scraped surface mixer.

drive shaft. At the same time, the ribbons on the end of the spokes impart a horizontal force on the product. The reduction in particulate damage is the key for products such as steak or chicken pot pies where maintaining identifiable meat pieces into the finished product is important for a consumer's perception of quality.

The vertical shaft agitator system such as the one in a hemispherical kettle is less effective because it sweeps the product in a circular motion around the agitator shaft but does not effectively lift the product vertically to counteract gravity. Manufacturers of hemispherical kettles have recognized the inefficiency of the vertical shaft agitator and have introduced kettles with sloped agitator shafts. Whereas this improves the mixing, particulates can also settle to the base of a slurry product, which can result in variable drain weights in finished product as well as overcooked pieces.

In some cases a second agitator is used with a type of propeller on the end which imparts a vertical force on the product when rotated at RPM. In all cases, these modifications to the basic vertical shaft agitator design do improve the performance of the agitation system but do not completely overcome the basic design limitation of the vertical agitator system.

Cooking vessels with horizontal agitation systems can be made with a single horizontal agitator or two horizontal agitators positioned side by side. Typically, the dual agitator mixer cookers are designed for cooking very viscous products. In some cases the agitators are designed so that the ribbons or paddles on the agitator shafts overlap each other. This type of "intermeshing" agitator is very effective in mixing the most viscous products because the overlapping of the rotating agitators keeps the product from rotating with either of the agitators. Effective mixing is enhanced since the product must intermix between the two agitators.

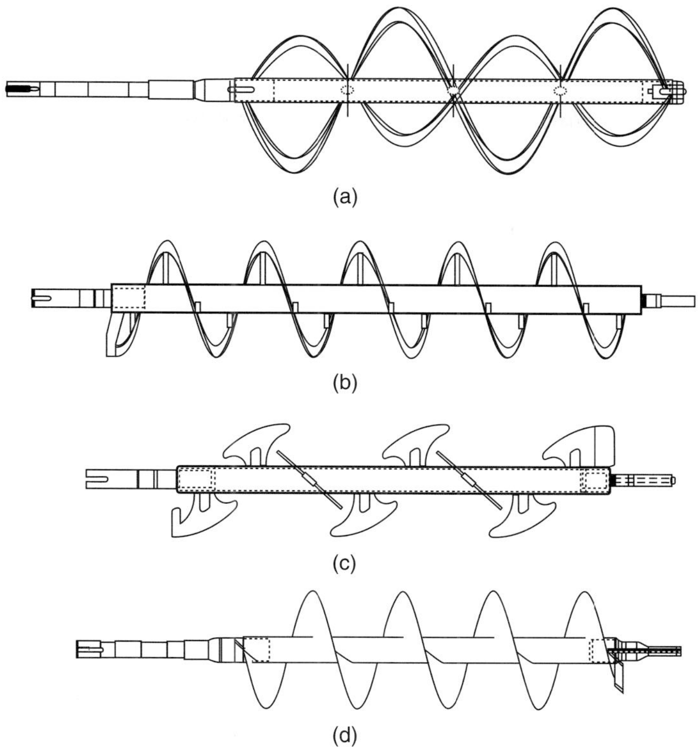

Figure 4.4. Agitators commonly used in horizontal mixer cookers (a) inner outer ribbon mixer, (b) simple ribbon mixer, (c) sickle paddle mixer, and (d) solid flight mixer.

A range of mixer styles can be used in horizontal mixers. Ribbon mixers are used for folding products where medium-to-low shear mixing is needed. Solid flight agitators are used when higher shear is needed and paddle agitators are used for the most viscous products such as chilled poultry or pet food slurries (see Fig. 4.4).

The RPM of the agitators will affect the quality of mixing; however, with most products it is not recommended to run the agitators at high RPM because it will damage the product. Products with fragile particulates can be damaged if the RPM is too high and some products such as cheese will separate if the product is worked too vigorously. The requirement of high rotary speed of the vertical shaft hemispherical kettle agitators to get adequate mixing is not desirable for these reasons.

Other Factors Affecting Heat Transfer

Heat transfer into the product is affected by the surface condition and finish of the cooking vessel. The most significant factor is the surface area of the cooking vessel that is covered by the heat exchange jacket. The more jacket surface area there is per cubic foot of product, the more heat energy will be transferred into the product and the faster the product will heat up. A simple way to compare the heat transfer performance of different cooking vessels based on jacket area is to divide the jacket area by the batch volume for each vessel. A horizontal agitator cooking vessel that extends the jacket area above the agitator shaft by wrapping the jacket and cooker wall around the agitator (as shown in the photograph later) will dramatically increase jacket area relative to the batch capacity of the vessel. However, it is not advisable to extend the jacket vertically above the agitator shaft to gain jacket area since the wall of cooker covered by the heat exchange jacket cannot be scraped with scrapers mounted on the agitator. These surfaces can yield burn on and produce negative off flavors.

Heat exchange efficiency can be improved by finishing the inside surface of the heat exchange area of the vessel. A smoother surface will transfer heat energy into the product slightly faster than a rough surface for two reasons. The first reason is that there is a smaller static layer of product between the food product and the heat exchange surface on a smooth surface. This results in better heat transfer. The second reason is that the scraper system will wipe off the product on the smooth surface more cleanly with each revolution of the agitator carrying the heated food particles away to be replaced with cooler particles. Since heat transfers into cool particles faster than into heated particles, the overall heat transfer is improved.

Another way to keep the heat exchange area clear of product as the agitator rotates is to coat the surface of the heat exchange area with a quick release material such as Teflon. Since the product slides off of such a surface more readily it is easier for the agitators and the scrapers to keep the heat exchange surface clear of heated particles of product. But there are considerations, with coatings such as Teflon, the coating can wear off and gets into the product. If a coating is used, it is important that it be selected carefully and maintained well to avoid any of the surface material flaking off into the product. Use of scrapers and a coating usually wears the coating faster because of the added abrasion by the scraper. Newer coating technology is available with greater adherence to the cooker wall but application in traditional cooking equipment remains limited and warrants further exploration and study (see Fig. 4.5).

Figure 4.5. The heated jacket area of a cooker relates directly to increased heat transfer. The above cooker has a wrap-around jacket which maximizes the heat exchange area.

The Purpose of Scrapers to Enhance Heat Transfer

It is important for the mixing system to keep the heated particles moving away from the heat exchange surface to improve heat transfer. It is also important to do this to minimize product burn-on. A high-quality scraping system does improve heat transfer but not necessarily for the obvious reasons. If a thin layer of product is allowed to rest on the heat exchange surface too long it will overheat and burn onto the heat exchange surface. Improving the mixing with effective agitators will minimize this thin layer of product called the "boundary layer," but effective agitation cannot eliminate it altogether. In addition to eventually causing burn-on, this boundary layer also insulates the heat exchange surface impeding heat transfer. When burn-on is permitted to form, it acts as an insulating layer, further slowing heat transfer.

The function of an effective scraping system is to disrupt the product in the boundary layer by wiping it off much like a squeegee wipes water off of a window. Although it does not completely eliminate the boundary layer, this wiping action forces this stationary product away from the wall so that it is picked up by the moving product being pushed around by the agitators. The scraper system must sweep the heat exchange surfaces

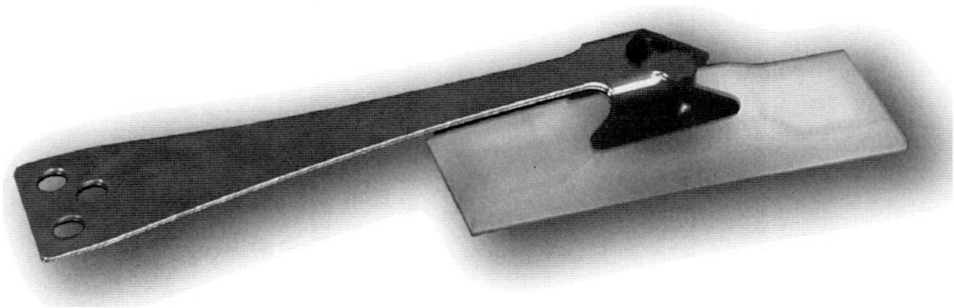

Figure 4.6. A spring-mounted scraper system with a pivot-mounted nylon scraper.

clean frequently, but it is the agitator that must distribute this heated product within the batch.

An effective scraper system has several characteristics. The most important of these is that the scraper must be spring mounted. Since the objective is to wipe away the boundary layer of product from the heat exchange surfaces, it is important to continuously press the scraper against the cooker walls with adequate force. Most hemispherical kettles rely on the product flow over the scraper to hold the scraper against the cooker wall. For smaller diameter kettles with low radial velocity, product flow over the scraper does not provide an adequate and consistent force to assure that the product resting against the wall is wiped away on a consistent basis.

It is not possible to fabricate an industrial size cooking vessel that is perfectly shaped. As a result there are always high spots and low spots around the cooker walls. For this reason an effective scraper system must pivot and rotate to maintain positive contact with the cooker wall as the agitator rotates. Many hemispherical kettles accomplish this by having the scrapers segmented so that the scraper is divided up into many narrow sections. The challenge with this design is that the spring system must exert the same spring force on each of these segmented sections to keep them firmly in contact everywhere around the cooker walls.

A more effective design is to use a scraper head that pivots (Fig. 4.6) on the end of the leaf spring that holds the scraper head against the cooker wall. By pivoting the scraper in the center, the scraper will always be flat against the cooker wall and the leaf spring will maintain firm contact with the full length of the scraper's leading edge as the agitator rotates.

The angle of the leading edge of the scraper is important. Ideally, the scraper blade would have line contact with the cooker wall so that the spring force is translated into a high force per square inch of contact area. With the scraper shown in Fig. 4.7, the contact surface of the scraper is

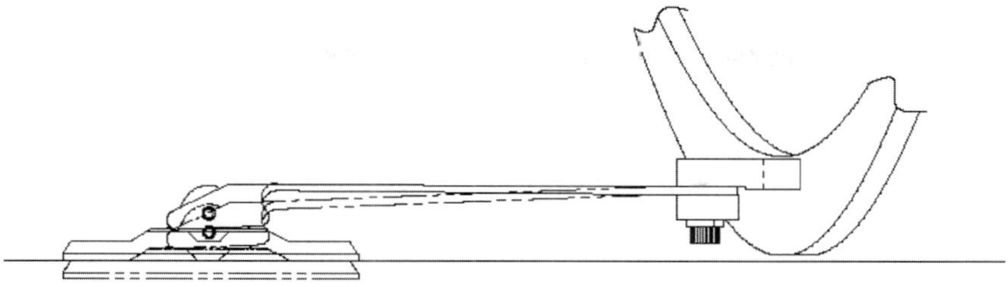

(a)

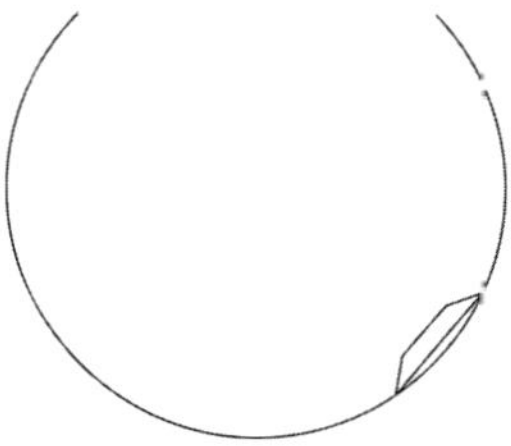

(b)

Figure 4.7. Leaf spring scraper allows float or flex to maintain contact with cooker walls which may vary (b). (a) A flat scraper pressed against a curved trough maintains line contact in either direction of rotation.

flat; however, the cooking vessel wall is curved. The flat shape against the curved surface of a cooker wall results in line contact with the wall, one on the leading edge of the scraper and another on the trailing edge. It is important that these two edges of the scraper remain sharp so that the scraper cuts into the product resting on the walls of the cooker rather than sliding over the boundary layer of product along the cooker wall.

The Importance of Different Types of Heat Sources

Effective heat transfer through a cooking vessel jacket is dependent on the heating medium inside of the jacket. Generally, the heating medium is hot water, steam, and thermal oil. These three heating mediums are used for three different heating results.

Cooking with hot water, as the heating medium, is done to minimize scorching or burning of the product. As an example, if the processor wants to melt chocolate, steam heat is too hot and will scorch the sensitive flavors in the chocolate. Hot water can be controlled very accurately at the best temperature for this type of application and the heat release rate into the product is very controllable. Additionally, hot water can be used temperatures equivalent to steam temperatures, if pressurized in the jacket, with less scorching and burning than with steam.

Steam is more universally employed for soups, sauces, and stew-type products. The rate of energy release from steam is higher but is also more difficult to control. There can be minor hot spots and cold spots from one part of the heated jacket to another part. Hot and cold spots can be substantial of the steam jacket which is not designed for even distribution of the steam. For this reason, many food processors seek equipment suppliers who are able to tailor the cooking equipment to achieve maximum cooking performance.

Although there are many reasons steam is so frequently used. One advantage is that steam boilers are inexpensive and the steam can be produced from natural gas or electricity. Steam is easy to pipe to different locations in the production plant and is by far the most popular method of cooking.

Another advantage of steam heat is that when a jacketed vessel is properly designed, the temperature at all parts of the steam jacket is nearly the same. Unless steam is superheated, the absolute temperature in the jacket is highly predictable and based on the pressure. Jackets heated with liquid such as water or thermal oil must be pumped through a specially designed jacket in a first-in-first-out flow pattern. The liquid is the hottest at the inlet to the jacket and from there, through the serpentine jacket loop, the heated liquid gives up heat to the product and cools down. The temperature at the discharge of the liquid-heated jacket is 10–15°F cooler than the liquid temperature going into the jacket. Additionally, liquid circulation pumps and the jacket serpentine path must be appropriately designed to ensure turbulent flow and in turn high heat transfer in the jacket.

The advantages of steam heat extend further. Steam gives off a great deal of heat energy as it condenses from gaseous to liquid. Although variable by pressure, the "latent heat of evaporation" of water is approximately 1,000 British thermal units (BTUs) per pound of steam. In other words, a pound of steam gives off 1,000 BTUs as it is condensed into a pound of water. As a simple comparison, 1,000 BTUs are enough to heat five gallons of water from 60 to 84°F. Steam also has an efficient heat transfer rate to the stainless steel wall of cooker. As the steam condenses on the inside of the cooker walls it beads up keeping the inside of the cooker wall moist. This moist surface transmits heat directly into the stainless steel wall of the jacket and on into the product in the cooker making it more efficient.

Thermal oil is used as a heating medium when the process requires very high surface temperatures. The ultra-hot surface temperature is required for a variety of ethnic dishes such as Chinese products or deeply browned meats. Egg-fried rice is a classic product requiring frying with thermal oil being pumped through the vessel jacket. The heat exchange surface must be hot enough so that the starch in the rice will not burn onto the walls of the cooking vessel. Temperatures up to 260°C (500°F) will burn rapidly whereas a temperature of 304°C (580°F) to 321°C (610°F) will cause the moisture in the surface of the rice to evaporate so quickly that the rice kernels literally jump up and down on the frying surface and will not stick. Many vegetable products cooked stir fried on a thermal oil-heated surface cause some browning and flavor development while also inactivating enzymes. For these products, cooking times are so rapid that they leave the texture of the vegetable fresh and crunchy.

Use of Heat Transfer Principles in Cooking Different Products

Cooking of different food products requires the application of these universal heat transfer principles differently. Cooking of meat products such as ground beef is different from cooking dairy-based products. Cooking products with heavy particulates is different from cooking products which are thickened with starch or pectin. Figure 4.8 illustrates a

Figure 4.8. Vegetable soups are a nonviscous liquid with lower density than the particulates which tend to settle rapidly.

chicken-noodle-type soup product in a single agitator mixer cooker. When high-density particulates are suspended in a low viscosity both, they tend to rapidly separate.

Dairy-based products must be cooked at lower steam pressures to reduce or eliminate burn-on since the milk or cream in these products can be very sensitive to high heated surface temperatures. Dairy-based products also require vigorous agitation to keep the product heating evenly and to scrape the walls of the cooker. This prevents the product close to the heat exchange surface from overheating because it stays against the steam jacket walls too long.

Cooking Meat Products

Cooking ground beef products such as taco meat is a unique process. It cannot be practically cooked with a steam temperature above 149°C (300°F) or 45 PSI steam. If the cooking vessel is operated at jacket temperatures higher than 300°F, the protein in the meat will burn onto the jacket walls.

The objective of cooking meat products will vary with each application; however, it is always necessary to make sure that the meat is completely cooked, safe, and that all pieces of meat and areas of the meat batch have been heated to the required temperature to kill pathogens and coliforms. This requirement in turn places emphasis on efficient mixing of the product during cooking and accurate temperature measurement. If the cooking system does not have an efficient mixing system, it is necessary to overcook major portions of the batch to ensure the coldest areas of the meat batch reach the minimum process temperature. These hot spots in the batch may have reduced texture, flavor, or viscosity. Although it may be safe to consumer, overcooked products often have reduced nutritive value and the processor may not wish to sell the product due to reduced quality. If the mixing system is efficient, the heat energy is evenly distributed during the heating process, and all the areas of the batch reach the required temperatures at the same time.

Without proper mixing pockets of under- or overcooked products may develop, but it is the key that the cooking vessel be equipped for proper detection and recording of the temperature. Most cooking vessels are equipped with some temperature-sensing device which is used to either modulate or isolate the heating source. Hemispherical kettles are often equipped with surface-mount thermocouples or RTDs. Although many of these designs are highly cleanable and are aesthetically pleasing, when installed near the jacket surface, they measure the temperature of product within the boundary layer or product immediately swept away from the jacket. This can be much higher than the bulk or internal temperature of the mixer. As a result, they are a poor choice for products which are difficult to mix.

Probe RTDs installed directly or within a cleanable thermowell are preferable for high viscosity products. Because they are installed so that they can more deeply penetrate the slurry, they give a more accurate reading of the average temperature. For well-mixed, horizontally agitated vessels, temperatures observed by the probe-type RTD tend to be accurate unless large particulates are used in the formula. The probe-type temperature measurement is not a substitute to manual validation and measurement but is an improvement over other alternatives.

When large particulates are used such as in steak pie filling, equilibration times should be determined for each recipe to ensure adequate cooking temperatures within the core of each piece. If the processor strictly regulates the incoming temperature and size tolerances for the meat pieces then the hold time extensions will be repeatable. Installation of separation screens at the discharge valve can be used to isolate any pieces excessive in size from continuing into the product flow. For example, if a process is established for a beef piece of 1/2 inch cube, then process hold time can be established for a 3/4-inch cube and a 3/4-inch separation device can be installed in discharge valve to prevent the passage of the larger potentially undercooked pieces. Equilibration times are best established through experimentation and hand measurement of the inside of the meat pieces.

Automation that is available with modern cooking equipment can be used to validate and ensure the proper process is being met. A chart recorder is commonly installed on a batch cooker to maintain a legal record that appropriate cooking times are used. With the widespread availability of PLC control system, measurement, validation, and automation can be digitized and used to ensure proper cook cycles are employed.

Most PLC control programs come equipped with password-protected automatic recipe control. After an appropriate cook cycle has been validated, it can be programmed into the PLC which will limit the completion of the batch until the approved cycle is completed. The use of PLC automation does not, in most cases, eliminate the need for a skilled operator but does help to ensure reliable, quality safe food is produced.

When temperature probes are used for process validation, it is important that they are calibrated with frequency to ensure accurate temperatures are observed and recorded. Calibration can be accomplished using boiling water at 100°C (212°F) or a water and ice mixture at 0°C (32°F). Probes should be calibrated at temperatures close to their intended operating conditions and with sufficient frequency to ensure safe production of product.

Affect of the Condition of Raw Meat in the Cooking Process

The condition of the meat product prior to cooking must be taken into consideration. Has the meat product been ground, diced, or emulsified? What is the fat content of the product? Has the product been frozen prior to cooking? What is the liquid content of the product? The answer to each of these questions will have a different affect on the cooking of a meat product.

If the meat product is emulsified, the protein in the meat can be separated easily during cooking and the protein will be cooked quickly onto the heat exchange surfaces. When cooking an emulsified product or a meat product with a percentage of emulsified meat in it, the pressure of the steam jacket must be low to avoid significant burn-on.

As previously discussed, ground meats are a popular product to cook due to their relative cost. Many slurry products such as taco meat, chili meat, meat, and gravy are substantially made of ground meat. How the ground meat is preprocessed or prepared is important. If the ground meat is ground too warm or mixed above 36°F, protein will be extracted in the grinding or mixing. This protein will burn onto the walls of the cooker and peel off during cooking and look and act much like rubber bands in the cooked product. Simply reducing the temperature of the meat to below 36°F during the grinding and mixing process will eliminate this cooking problem. Even when the ground beef has been ground properly; it must be cooked with a jacket temperature below 154.4°C (310°F), which is approximately 45 PSI steam pressure. If it is cooked above this temperature, beef will burn on and insulate the heat exchange surface even when cooked with spring-mounted scrapers and a horizontal agitator system vessel.

Ground meats with a great deal or soluble protein will stick to some types of polymer materials used for scraping devices. The most common plastic for the manufacture of scrapers is injection-molded nylon which is hydroscopic, meaning it absorbs water. The soluble protein in the meat mixture will be attracted to the scraper causing it to accumulate on the scraper surfaces. For products such as these, changing to a different type of plastic in the scraper material that is not hydroscopic will solve this problem. The equipment manufacturer should make a range of materials available for scrapers to accommodate processing conditions such as high temperature, low pH, and high protein.

Cooking ground meat that was frozen prior to grinding will take much longer in the cooking vessel—as much as 50% longer to cook. The reason for this may not be obvious to the operator. Processors typically will thaw the frozen meat to about −1 to 0°C (30–32°F) and then grind it. The heat generated during the grinding process will raise the temperature of the meat a couple of degrees so they think the meat is completely thawed because it looks thawed. However, there are cells in the meat that are still frozen and these cells cause problems during the cooking process. If other frozen ingredients are used in a formula, a dual agitator mixer cooker can be employed to ensure frozen pieces to not "block" freeze together preventing even, safe cooking.

Ground meats are also cooked gradually in a large cooking vessel. Even in a horizontal cooker, some of the meat will heat up faster than other parts of the meat mixer. This heated meat turns into a cooked slurry whereas the colder uncooked meat rolls into balls, some as large as a small melon. These balls are cooked from the outside. The meat on the outside of a ball is cooked and peeled off making the ball of uncooked meat smaller. The process of cooking these balls of meat takes time. If the ball is made up of frozen cells of meat it takes much longer to cook from the outside and thus the whole batch takes longer to cook completely eliminating the uncooked spots in the batch. As with frozen products, the use of a dual agitator cooking device helps to break apart the pieces ensuring an adequate thermal process. Although not a substitute for proper cooking, a separation screen, hot hold times and scraped surface can help to avoid improperly processed meats from continuing into the process flow.

If the recipe for the ground meat product that is being cooked has other liquid ingredients, the liquid will absorb heat from the heat exchange jacket and the whole batch will heat up and cook faster. The added liquid may reduce the risk of burn-on and therefore will allow cooking with a higher steam pressure. On the other hand, if the added liquid is a milk-based ingredient or an ingredient with high amounts of sugar in it, the risk of burn-on is much higher and the steam pressure in the jacket must be lowered to keep the jacket surface temperature lower.

Cooking Dairy-Based Products or Products with High Sugar Content

There are a whole new set of processing considerations when cooking products with milk or cream or products high in sugar. Milk-based products tend to boil very easily and foam when the product gets too hot causing the batch of product to boil over. If the heat exchange surface is too hot the milk in the product will scorch and take on an off flavor. Fruit products high in sugar content will also foam readily and will burn-on easily as well. With these types of products the basic factors that control burn-on must be followed closely:

- Select a cooking vessel that has an effective mixing system.
- Make sure the scrapers are spring mounted and have a sharp leading edge.
- Lower the temperature of the heat exchange surface.
- Speed up the RPM of the agitators a little.
- If the product starts to foam and the surface of the batch begins to rise, turn down the temperature of the heat exchange surface even further.

Cooking Products with Particulates

Cooking products with a light, nonviscous sauce with heavy particulates that sink or float carries another set of problems. It is important to keep the particulates in suspension. Some vegetable ingredients tend to float whereas other ingredients such as pasta tend to sink rapidly. A properly designed horizontal style agitator is the best choice for this type of product since the action of the agitators is more effective in agitating the batch vertically. The agitator forces provided by the agitator rotation push the floating pieces down and lift the sinking pieces up keeping the batch evenly in suspension without requiring excessive agitator RPM that might cause particulate damage.

Many agitator styles can be used in a horizontally mixed cooker as previously discussed; however, with products that have a tendency to settle or float, selection of the ideal mixer should be carefully considered.

Cooking a product, that is thickened with modified or natural starch, places additional emphasis on effective agitation. If the granules are not heated to the proper gelatinization temperature, they do not swell and therefore do not thicken the product. If they are overheated, they burst and the absorbed water is released, so the thickening characteristics of the starch are lost. The starch must be heated to the right temperature to bloom completely and to get the full benefit out of the cooking process.

Since vertical agitator cooking vessels do not mix well and do not heat the product evenly, they do not bloom starch well and as a result the processor must increase the starch percentage to get the viscosity of product they want. With a horizontal agitator system, because it is an excellent mixer, it is possible to actually reduce the starch percentage because a higher percentage of starch that is added will bloom properly.

Challenges of Chilling Slurries

Introduction

There are a number of different methods of chilling a slurry product. The slurry can be pumped through a tube in tube heat exchanger or inline scraped surface heat exchange. Alternatively, a cryogenic can be added such as liquid nitrogen, liquid carbon dioxide (CO_2), or dry ice (solid CO_2). The addition of a cryogenic is quite expensive. The product can be placed into shallow trays and chilled in a blast freezer. This method is labor intensive and results in substantial product loss through product spills and crusting over of the product due to drying of the surface.

Regardless of the method of chilling it is important to ensure that the food product is rapidly chilled to 5°C (40°F) to prevent the growth of spoilage bacteria, pathogenic bacteria or formation of toxins. For food service, the warmest point in the food should be chilled between 57°C (135°F) and 21°C (70°F) in less than 2 hours and from 21°C (70°F) to 5°C (40°F) in less than 4 hours.

An efficient method of chilling is to chill the product in the vessel in which it was cooked. This is often referred to as cook-chill. Chilling of slurries in the cooking vessel has many of the same challenges as heating slurries. A common target is to complete the chilling process in less than 90 minutes. With a properly designed process and appropriate equipment, it is often possible to cool even faster.

Unfortunately, the heat transfer of heat energy out of a product is significantly slower since the temperature difference (ΔT) is much less when chilling. Remember, the rate of heat transfer is directly related to the "ΔT" temperature difference. In heating a product the "ΔT" is more than 93°C (200°F), whereas the "ΔT" in chilling is usually half as much; therefore, the factors that we have discussed previously are twice as important.

Jacket Chilling Strengths and Weaknesses

When chilling a slurry product by circulating a chilled liquid through the steam jacket, having effective mixing of the product is most important. Keeping the warmest product in constant contact with the cold energy transfer surface is the key to chilling a product with a chilled water or chilled glycol jacket. An effective scraper system is also important because a clean energy exchange surface enhances jacket cooling. Without scrapers, a coating of product resting next to the energy exchange surface insulating the jacket and reducing heat transfer from the product. On Blentech cookers, jacket cooling rates have been increased more than 15% when scrapers are used.

Even with an effective agitation and scraping system the rate of cooling can be inadequate to achieve the required rates of cooling with the same vessel used for cooking. In most high viscosity products, chilling with 4 hours cannot be accomplished using jacket chilling without external heat exchangers. The heat transfer is too slow.

Figure 4.9 illustrates this dilemma. It shows a typical rate of cooling for a viscous cheese and on sauce. In this example, the cheese and onion sauce is chilled with 1°C (33°F) water through the jacket. This product

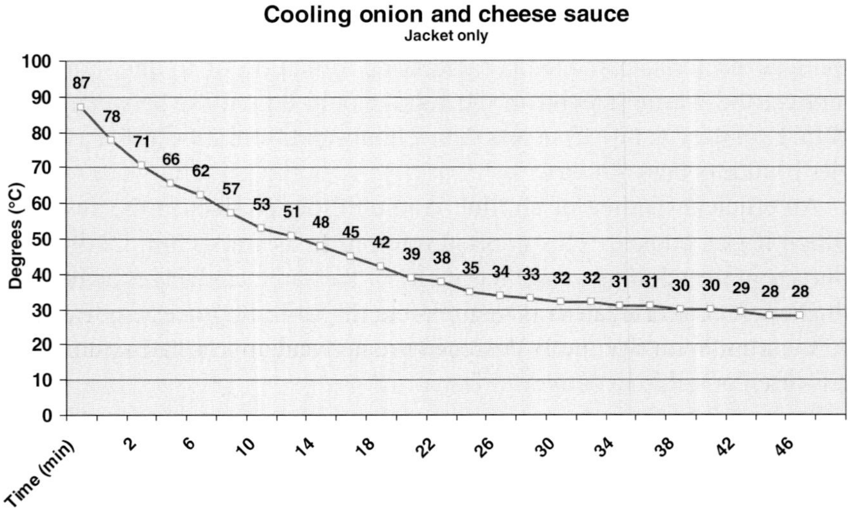

Figure 4.9. Rate of jacket chilling is limited by low temperature differential between product and jacket.

which thickened as it cooled, formed a substantial boundary layer and limits heat transfer. Practically speaking, the rate of cooling declines so rapidly that jacket cooling is not capable of chilling this product below 20°C (71°F) which could not meet the cooling standard for cook chill processes. At 28°C the chilling rate was down to a fraction of a degree per minute. It would have taken many hours to chill the product to 5°C (40°F). In addition, agitating the cooked product for a many hours would have destroyed any particulates in the product.

Vacuum Chilling—Solution or Problem

An interesting method of chilling products is vacuum cooling in the cooking vessel. This is a chilling process that not commonly used. Basically, a vacuum is pulled in the head space above the batch of product. This vacuum level is dropped below the vapor pressure of the water in the product causing the water to evaporate rapidly or boil. The deeper the vacuum the faster the product boils and the lower the temperature of the product. Most products can be cooled to 5°C (40°F) in around 30 minutes. The reason for this rapid cooling is that it requires a substantial amount of heat energy to change water from a liquid to a gas (approximately 1,000 BTUs/lb of evaporated water). As a vacuum is increased the heat energy is released by the evaporation of water out of the product. The process is simply to sustain boiling at lower and lower temperatures until chilling is achieved.

There are a few practical limitations as to how fast a slurry product can be cooled using vacuum cooling: foaming tendency and splattering. Because the product boils violently if the vacuum depth of the vacuum is increased too rapidly, there is the risk that the product will erupt and splatter so violently that a substantial amount of the product can be extracted into the vacuum generator machine. This product cannot be used thus reducing yields and it can damage the vacuum-generating machine. Because this is a common risk, most processors using vacuum cooling systems design an inspectable, CIPable system including all of the interconnecting pipes.

A common concern for producers with very little vacuum cooling or cooking experience is that vacuum cooling will extract volatile aroma components reducing the flavor of the product. While some volatile compounds are extracted, for many products, losses are not significant and a high quality, high flavor product results. Nevertheless, vacuum cooling

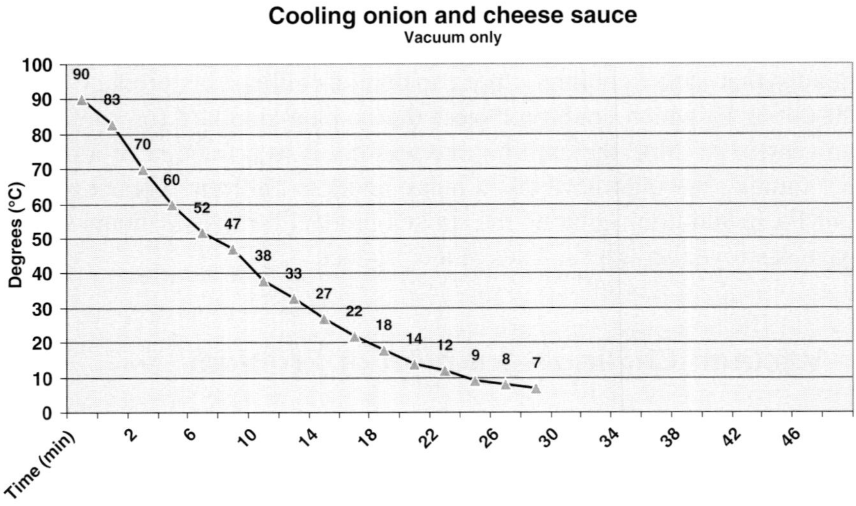

Figure 4.10. Rate of vacuum chilling is limited only by the rate of boiling that can be achieved without foaming.

must be carefully controlled especially when the product is hot when the vacuum cooling process is first started.

The primary benefit offered by vacuum cooling process is rate of cooling. Figure 4.10 shows the rate of vacuum cooling of the same cheese and onion sauce formulation chilled in Fig. 4.9. This illustration shows the vacuum cooling of the same sauce cooled by jacket cooling can be accomplished in significantly lower time and is less limited in its final temperature. With a properly designed vacuum cooling system, it is practical to chill the product down to 5°C (40°F).

ComboChill System (Patented)—the Best of Both Worlds

A patented system of chilling called the ComboChill system by Blentech Corporation is designed to combine jacket chilling with vacuum chilling in a very unique way. In this way, flavors are preserved while ensuring that the food is chilled rapidly enough to maintain a safety and preserve shelf life. Since jacket chilling is almost as efficient as vacuum chilling at higher temperatures and does not extract volatile components, this method of chilling can be used between 190°F (88°C) down to about 105°F (41°C). As is shown by the slope of the jacket chilling curve, the

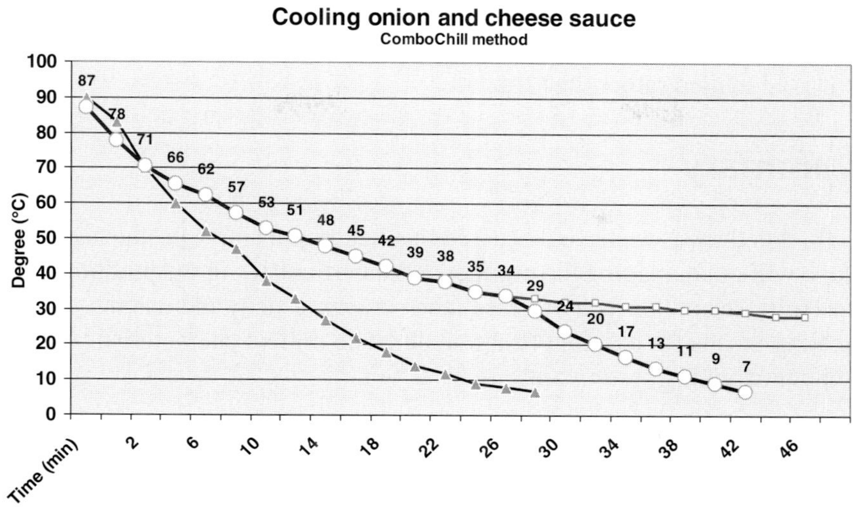

Figure 4.11. Patented ComboChill process of jacket and vacuum cooling meets cooling standards while preserving flavor in products.

cooling rate of the product is quite acceptable, cooling this product to 34°C (93°F) in only 28 minutes (Fig. 4.11). However, the cooling rate with the jacket cooling slows down significantly below 33°C (93°F).

Interestingly, the cooling at these higher temperatures is where vacuum cooling is volatile and unstable. Below 38°C (100°F) vacuum cooling continues to be a rapid method of cooling slurry products but it extracts far less volatile compounds. At the lower temperatures the flavor volatiles are more stable and tend to remain. Additionally, the boiling rate is lessened so there are fewer problems around having the product pulled into the vacuum system. Finally, particulates are not damaged at lower temperatures as they are with vacuum cooling at higher temperatures.

The patented ComboChill system uses jacket cooling where it is efficient and vacuum cooling where it efficient and avoids the temperature range where each is not. The solution is simple and yet effective. With the ComboChill system it is practical to chill in the same vessel in which the product is cooked and accomplish the cooling to 40°F (5°C) in much less than required by the proper food safety regulations.

Since the product does not have to be pumped to a separate vessel, the type of pumping and shear damage observed in an inline cooling system does not occur. After products are chilled, their viscosity increases so that the particulates are easily held in suspension and maintain consistent drain weight through the filler into the container. This method of chilling products for refrigerated distribution has become extremely popular in the European markets where a high percentage of ready meals are distributed and sold chilled rather than frozen.

Summary

In this review, we have discussed that heating of slurry products can be done without damaging the product or the particulates in the product if we carefully select the type of vessel and scraper system and use the proper heating means for efficient heat transfer. The critical characteristics of an efficient cooking system are:

Method of Heating

- For most applications, pressurized steam is the most efficient heating means for most cooking applications.
- Hot water is effective when heating highly heat sensitive products such as dairy- or sugar-based products.
- Thermal oil can be utilized if the application requires very high heated surfaces for special ethnic products.

Distribution of the Heat Energy into the Product

- Heat transfer into the slurry is vastly improved if the cooking vessel has an efficient mixing system—preferably a horizontal agitator or dual agitators.
- To avoid overheating a slurry product, the mixing system must be optimized (design, RPM, and scraping).
- Scrapers do not keep the product from burning on but rather wipe heat exchange surfaces to push the boundary layer of product away from the cooker walls enabling the agitator to mix the heated product with the remainder of the batch.

Cooking Various Products and Applications

- Aggressive mixing systems (RPM) can damage fragile particulates and lower product quality (diced vegetables and diced potato).
- Preprocessing of meat products (grinding/mixing) at temperatures above 2°C (36°F) can increase burn-on.

- Too high a jacket temperature (steam pressure) can slow up the cooking process by increasing burn-on (ground beef or dairy-based products).
- Vacuum cooking can improve product quality and flavor.

Chilling Products Can Be Efficient in the Cooking Vessel

- Jacket cooling is slow below 100°F (38°C).
- Vacuum cooling is volatile and unstable above 100°F (38°C).
- Combining jacket chilling and vacuum chilling will avoid the disadvantages of both systems.

Reference

Warriss, P.D. 2000. *Meat Science: An Introductory Text*. CABI Publishing, Bristol, UK.

CHAPTER 5

Processing Interventions to Inhibit *Listeria monocytogenes* Growth in Ready-to-Eat Meat Products

C. Lynn Knipe, The Ohio State University

Listeria monocytogenes is commonly found throughout nature, and can survive for long time periods in low oxygen and refrigeration temperature environments. With consumer demands for reduced salt products, and the increase in national, wholesale distribution systems, which require that products have longer refrigerated, shelf lives, the opportunity for *L. monocytogenes* to survive and grow, in vacuum-packaged, ready-to-eat (RTE) meat products has increased. This is particularly a problem with the potential for environmental contamination of products between the cooking and packaging steps (e.g., peeling, slicing, and other handling during packaging processes). As a result, additional steps, or interventions, need to be taken to inhibit, or prevent, *L. monocytogenes* from surviving or growing on RTE meat products, particularly if these products are contaminated by the processing environment after cooking, but before packaging.

There are many processing interventions, some ingredients and some processes, which inhibit the growth of *L. monocytogenes* in RTE meat products. A number of these interventions are traditional methods that have been used by processors for many years; however, there are many new ingredients and processes that are being developed to destroy or inhibit *L. monocytogenes* growth.

According to the USDA-FSIS Compliance Guidelines (USDA-FSIS, 2006), antimicrobial interventions can be used both as postlethality treatments and growth inhibitors. Postlethality treatments are lethality treatments which are applied after and/or are effective after postlethality exposure to potential environmental contamination. As postlethality treatments, interventions need to be validated to achieve a minimum of 1-log reduction in *L. monocytogenes* compared to the same product that did not receive the intervention treatment. As growth inhibitors, interventions must be validated that they inhibit the growth of *L. monocytogenes* to not more than 2 logs of growth over the duration of the product's shelf life.

Traditional Antimicrobial Processes and Ingredients

Traditional processing methods that inhibit the growth of *L. monocytogenes* include freezing, using high salt levels, acidifying, and drying products. These antimicrobial processes can be used in place of the addition of antimicrobial ingredients to allow a product to be produced under either Alternatives 1 or 2, as described in the USDA-FSIS Compliance Guidelines for Listeria (USDA-FSIS, 2006).

Freezing and storing RTE meat products at temperatures below −0.4°C (31°F) prevent the growth of *L. monocytogenes*, but do not destroy the *L. monocytogenes*. This means that upon thawing the product, *L. monocytogenes* would continue to grow. Therefore, to use freezing as an antimicrobial process, products would have to be kept frozen throughout their distribution shelf life.

Meat products, to which 3% or more of sodium chloride are added, or in which during drying the salt content increases to exceed 3%, may not need to be considered for one of the processing alternatives (USDA-FSIS, 2003). However, sodium chloride levels typically used in RTE meat products would not be very effective in preventing the growth of *L. monocytogenes*.

Sodium chloride levels of 2 and 3% in ground pork showed significantly lower *L. monocytogenes* reduction levels (5.01 and 4.13 log CFU/g, respectively) compared to 0.5 and 1% salt levels (6.75 and 6.36 log CFU/g, respectively) (Yen et al., 1991). Addition of 3.5% sodium chloride (along with 200 ppm sodium nitrite and 300 ppm sodium nitrate) increased the heat resistance of *L. monocytogenes* when cooked to 55–70°C (131–158°F) (Mackey et al., 1990).

Models developed to predict the amount of lactate and diacetate needed in different cured and uncured meat products, showed that the growth rate of Listeria was decreased with the increased addition of sodium chloride in uncured meat products, where it was a minor factor in the cured product model (Seman, 2001). However, *L. monocytogenes* has been found to survive in chilling brine solutions that contained 10 and 20% sodium chloride (Valderrama et al., 2008).

Sodium nitrite contributes to the overall suppression of *L. monocytogenes* growth, but added to meat products, as the only antimicrobial agent, has not been shown to be effective in inhibiting *L. monocytogenes* growth (Doyle, 1999). It has been reported that addition of a minimum of 30 ppm sodium nitrite to a turkey product was needed to enhance the antilisterial activity of the combination of potassium lactate and sodium diacetate (Glass and McDonnell, 2008). The bacteriostatic effect of sodium nitrite on *L. monocytogenes* is enhanced by the anaerobic environment of vacuum packages and cold storage temperatures (Buchanan et al., 1989).

Seman's models, developed to predict the usage levels of lactate and diacetate needed in different meat products, indicated that nitrite was an important inhibitor of Listeria growth (Seman, 2001). Even with the use of lactate and diacetate, without sodium nitrite in a meat product, it was difficult to achieve zero Listeria growth (Seman, 2001).

The addition of nitrite (concentration not reported) and 3% sodium chloride increased the *D* values (increased heat resistance) for *L. monocytogenes* by a factor of 5–8 (Farber, 1989). Also, the addition of 200 ppm sodium nitrite and 300 ppm sodium nitrate (along with 3.5% sodium chloride) increased the heat resistance of *L. monocytogenes* when cooked to 55–70°C (131–158°F) (Mackey et al., 1990).

The addition of 0.4% sodium acid pyrophosphate (SAPP), along with 40 ppm sodium nitrite and 0.26% sorbate to frankfurter formulations, has been shown to extend the production of botulinal toxin and reduce the numbers of toxic samples, compared to frankfurters made without SAPP (Wagner and Busta, 1983), and SAPP inhibited toxin production more than nitrite and sorbate on pork and beef frankfurters, that were temperature abused at 27°C (81°F). It has also been claimed that a synergism existed between SAPP or sodium hexametaphosphate (SHMP) and sorbic acid in extending the time to botulinum toxin formation in canned ground pork (Ivey and Robach, 1978). SAPP has been shown to be the most effective inorganic phosphate (compared to SHMP and sodium tripolyphosphate) in inhibiting on *C. botulinum* growth and toxin production in chicken frankfurters that were stored at 27°C (81°F) (Nelson et al., 1980).

Other ingredients used in making RTE meat products may also increase the heat resistance of *L. monocytogenes* in these products. For example, researchers have reported that the addition of either 1% dextrose or 0.4% phosphate (0.2% sodium tripolyphosphate and 0.2% sodium SHMP) to ground pork significantly reduced the destruction of *L. monocytogenes* when cooked to 60°C (140°F), compared to the control that contained no nonmeat ingredients (Yen et al., 1991). However, 0.055% sodium erythorbate was found to have no protective effect on *L. monocytogenes* in a cooked product made from ground pork (Yen et al., 1991). A ground pork product made with a combination of 2% sodium chloride, 1% dextrose, 0.4% phosphate, 0.55% sodium erythorbate, and 0.0156% sodium nitrite resulted in a 3.8-log less reduction in *L. monocytogenes*, compared to the ground pork control, with no added ingredients (Yen et al., 1991).

Fermentation of meat products to reduce the pH of the product to below pH 4.4 would be considered an antimicrobial process. Drying of products may or may not be combined with the fermentation process, but if the products are dried to a water activity (A_W) below 0.92, the drying process would also be considered an antimicrobial process. An RTE product with an A_W of 0.85 or lower would be considered shelf stable. If fermentation and/or drying are listericidal, or cause a reduction in Listeria, they could also be considered as postlethality treatments.

Product composition can also impact the resistance of *L. monocytogenes* to thermal destruction. Fat in meat products has been shown to slightly increase the heat resistance of *L. monocytogenes* (Mackey et al., 1990).

Hurdle Technology

The antimicrobial ingredients and processes discussed in this chapter are sometimes studied and discussed individually, in an attempt to determine the impact of each ingredient or process on inhibition of *L. monocytogenes*. However, it is well known that combining multiple ingredients and processes (also referred to as hurdles to *L. monocytogenes* growth) will have greater impact on producing a safer product, while potentially allowing lower levels of antimicrobial agents which might otherwise negatively affect the product flavor. This is referred to as hurdle technology (Leistner, 2000).

Novel Antimicrobial Agents

Antimicrobial agents are available to prevent the growth of *L. monocytogenes* in RTE meat products. Presence of *L. monocytogenes* is typically not the result of inadequate cooking treatments, but postcook contamination of products, from *L. monocytogenes* in the environment. Ingredients must be validated for their effectiveness in preventing the growth of *L. monocytogenes* (by 1–2 logs through the duration of the shelf life of the product) to be used as antimicrobial agents, to satisfy requirements for either Alternative 1 or 2 (USDA-FSIS, 2003). These ingredients are considered bacteriostatic, because they prevent growth, but are not necessarily bactericidal, which would involve destroying the pathogen.

These antimicrobial agents are applied to the product in a variety of ways, such as: added to meat products as part of the product formulation, added to the surface of the finished product before packaging, or added to the packaging materials that are used for packaging RTE meat products. A list of antimicrobial ingredients can be found in the Table of Safe and Suitable Ingredients (USDA-FSIS, 2007); however, this list is constantly being updated, so you are encouraged to check this list regularly for new antimicrobial agent options.

Antimicrobial agents may include organic acids, bacteriocins, ozone, liquid smoke extracts, spices, etc.

Organic Acids

Organic acids which are most commonly added to meat products include acetic, citric, lactic, malic, etc. They may be added as acids or salts of the acids, and are added to meat products to reduce the pH of the water phase of the product, which results in inhibiting growth and/or death of microorganisms (Samelis and Sofos, 2000). Antimicrobial efficiency of organic acids has been shown to depend on pH, water activity, moisture, fat, nitrite, and salt content.

Dipping solutions of 2.5% acetic acid, 2.5% lactic acid, and 5% potassium benzoate had listericidal effects on cooked ham and bologna slices, when stored at 10°C (50°F); however, 0.5% solutions of nisin were not effective at controlling *L. monocytogenes* under the same conditions, unless combined with one of the three organic acids (Geornaras et al., 2005). Acetic acid (MOstatin V, WTI, Inc.), as well as acetic acid combined with lemon juice (MOstatin LV1) has been shown to allow less than 0.5 log

CFU/g growth of *L. monocytogenes* in uncured roast beef over 120 days of storage at 4°C (40°F) (Hofing et al., 2008).

Surface spraying of lactic acid onto bologna, ham, and miniature frankfurters significantly reduced *L. monocytogenes* levels by more than 1 log CFU/g, at time 0, reduced *L. monocytogenes* counts by more than 2 log CFU/g after 7 days of storage, and reduced *L. monocytogenes* counts to below detection levels after 90 days of storage at 4°C (40°F) (Ahmed et al., 2008). Other research involved applying an organic acid solution, containing acetic, lactic, benzoic, and propionic acids to the surface of RTE frankfurters, in combination with steam surface pasteurization, to control *L. monocytogenes* (Murphy et al., 2006).

Surface application of organic acids would reduce the need to add antimicrobial ingredients to the product formula, and pulling a vacuum during packaging of surface-treated products should improve the distribution of the organic acids in contact with the product surface.

Salts of organic acids such as sodium and/or potassium lactate and sodium diacetate are effective in inhibiting *L. monocytogenes* growth in RTE meat packages and may be added as antimicrobial agents to meat product formulations at maximum levels of 4.8% and 0.25%, respectively (USDA-FSIS, 2007). Lactates and diacetates may be added separately to prevent the growth of *L. monocytogenes*; however, a synergistic effect between lactates and diacetates in inhibiting the growth of *L. monocytogenes* in RTE meat products has been found, which allows lower levels of lactate and diacetate when they are added together to get the same bacteriostatic effect, as when added individually (Glass et al., 2002). The antilisterial effect of lactates and diacetates has been found in both vacuum packaged and aerobically stored RTE meat products, which means that these ingredients provide protection against Listeria growth after RTE meat packages are opened by consumers (Mbandi and Shelef, 2002). Lactates accomplish this bacteriostatic effect by increasing the lag phase of *L. monocytogenes* (Jofre et al., 2008).

Prior to its use as an antimicrobial ingredient, lactate was added to meat products to increase water-holding capacity and cooking yields (Duxbury, 1988; Evans et al., 1991; Shelef, 1994).

The effect of sodium lactate and sodium diacetate, added individually and in combination with glucona delta lactone (GDL) to bologna was evaluated with inoculated, vacuum packaged, bologna slices at two temperatures (4°C and 10°C, or 40°F and 50°F), over 90 and 28 days of storage, respectively (Barmpalia et al., 2005). The combination of 1.8% sodium lactate and 0.25% sodium diacetate was the most effective treatment in preventing *L. monocytogenes* growth at both temperatures.

The addition of 0.125% GDL along with the above levels of sodium lactate and diacetate was the second most effective treatment again at both storage temperatures. These results disagreed with earlier data, from the same laboratory, which showed that 0.25% GDL was equal to 0.25% sodium diacetate in preventing the growth of *L. monocytogenes*, when combined with 1.8% sodium lactate, and stored at 4°C (40°F) (Samelis et al., 2002). One explanation for this difference could have been that the earlier study used frankfurters as the test medium and the more current study used bologna slices; however, otherwise, both products were made using the same basic formula.

Lactates and diacetates are effective when added to product formulations, but not when applied to the surface of products as a dip (Glass et al., 2002; Mbandi and Shelef, 2002). Sodium lactate can be added up to 2–3% of product formula, without causing flavor problems; however, sodium lactate can cause a salty, metallic-type flavor, particularly in uncured products, if added at high levels (Seman, 2001). It has also been reported that at levels of 3% or higher, lactate increased a sour taste for cooked beef top rounds, and 4% lactate caused a mild throat irritation for some sensory panelists; however, fresh beef flavor notes increased and warmed-over flavors decreased with increasing levels of lactate (Papadopoulos et al., 1991b).

To the contrary, sodium lactate added to the formulation of beef top rounds, added at a 3% level, has been shown to enhance fresh beef flavor, minimize warmed-over flavor, resulting in a stronger, "beefy/meaty flavor" (Papadopoulos et al., 1991a).

Sodium diacetate can also be detrimental to flavor (e.g., vinegar-like) and aroma in products when added at levels of more than 0.12% (Stekelenberg and Kant-Muermans, 2001).

In the case of uncured and unsmoked products, higher levels of sodium lactate and diacetate are needed in RTE products, to inhibit *L. monocytogenes*, compared to cured and smoked products (Glass et al., 2002). In particular, uncured and unsmoked products such as turkey breasts and roast beef products are often mildly flavored, which would suggest the need for flavorings to mask the flavors of the higher levels of sodium lactate and diacetate. Addition of sodium chloride has also been shown to reduce the effectiveness of lactates (Chen and Shelef, 1992; Shelef and Yang, 1991).

The addition of lactate and diacetate to RTE meat products is more effective when these products are stored at 4°C or less (Barmpalia et al., 2005; Glass et al., 2002; Jofre et al., 2008).

The effects of lactate and diacetate on the inhibition of *L. monocytogenes* in varying product compositions have also been studied, using three types of meat raw materials, including pork trimmings, turkey breasts, and whole boneless hams (Seman et al., 2002). A predictive equation for determining the amount of lactate and diacetate needed in meat products was developed from this research. This program, the Opti.Form Listeria Control Model (Purac Company, www.opti-form.com), models the growth of *L. monocytogenes*, in both cured and uncured products, based on the finished product pH, salt, moisture content, and temperature.

Levels of lactates needed to inhibit *L. monocytogenes* increase with increased moisture contents and water activities in meat products (Chen and Shelef, 1992; Shelef and Yang, 1991).

Sodium and potassium lactates have also been shown to inhibit clostridia growth in cook-in-bag turkey products (Maas et al., 1989). Lactates have also been shown to inhibit *Clostridium sporogenes* growth in uncured beef roasts, that were cooked in cook-in bags, and temperature abused (Unda et al., 1991).

Regarding product quality, lactate has also been shown to reduce color fading of vacuum-packaged beef bologna (Brewer et al., 1992) and to cause precooked beef roasts to appear darker and redder, with less surface graying (Papadopoulos et al., 1991b).

Sodium lactate also increases salt flavor intensity and decreases off flavor development when added to beef bologna (Brewer et al., 1992), and has been shown to enhance fresh flavor notes, minimize warmed-over flavor notes, and result in stronger beefy/meaty flavors when added to precooked beef roasts (Papadopoulos et al., 1991a).

The addition of lactate to precooked beef rounds has also been shown to lower sensory scores for rancidity, reduce lipid oxidation values, and decrease flavor deterioration (Maca et al., 1999). Sodium lactate was also found to reduce oxidation of pork and reduce TBARS (thiobarbituric acid reacting substances) formation nearly as well as BHT, when the pork was stored at 0–5°C (32–41°F) (Nnanna et al., 1994).

Lactates have also been shown to improve cooking yields of meat products. Cooking yields of beef top rounds (Papadopoulos et al., 1991b) and other precooked beef rounds (Maca et al., 1999) increased with increasing the lactate level.

Cooked smoked sausage, which contained 1.5% potassium lactate and 0.05% sodium diacetate, was immersed in a variety of antimicrobial agents to determine the effect on *L. monocytogenes* growth (Geornaras et al., 2006). The postprocessing immersion treatment of either 2.5% acetic acid, 2.5% lactic acid, 5% potassium benzoate, or 0.5% nisin (Nisaplin) resulted in an initial reduction in *L. monocytogenes* that was not affected by the presence or absence of lactate and diacetate. Immersion solutions which contained mixtures of nisin and the organic acids were much more effective in the initial reductions in *L. monocytogenes* than that of the organic acid solutions alone. Similar initial reductions in *L. monocytogenes* were found when bologna and ham were immersed in the same antimicrobial agents (Geornaras et al., 2005).

Immersion of smoked sausage in a variety of antimicrobial agents alone, or in combination with nisin alone, was not effective in preventing the growth of *L. monocytogenes* when stored at 10°C (50°F) for 48 days (Geornaras et al., 2006). The antimicrobial solutions tested included 2.5% acetic acid, 2.5% lactic acid, 5% potassium benzoate, or 0.5% nisin (Nisaplin). However, these researchers found that immersion in the same antimicrobial agents was effective in preventing *L. monocytogenes* growth in bologna and ham slices (Geornaras et al., 2005). It is believed that the difference in results is due to the increased absorption of the antimicrobial agents by the bologna and ham slices, compared to the smoked sausage, during the immersion process.

In some studies, buffered sodium citrate has been shown to increase *L. monocytogenes* growth (Stekelenberg and Kant-Muermans, 2001), or to have no effect in preventing *L. monocytogenes* growth in RTE ham products (Poovey et al., 2008). Buffered sodium citrate significantly reduced cooking yields of RTE hams (Poovey et al., 2008).

On the other hand, buffered sodium citrate and sodium diacetate have also been proven to inhibit *L. monocytogenes* growth in RTE meat products (Poovey et al., 2008). Buffered sodium citrate and sodium lactate (IONAL LC, WTI, Inc.) have been shown to allow less than 0.5 log CFU/g growth of *L. monocytogenes* in roast beef over 120 days of storage at 4°C (40°F) (Hofing et al., 2008).

In addition, the combination of buffered sodium citrate and sodium diacetate has been shown to inhibit *C. perfringens* germination and outgrowth during postcook chilling (Thippareddi et al., 2003). This research showed that cooling times for roast beef and pork loins (from 54.4 to 7.2°C, or 130 to 45°F) could be safely extended to 21 hours, with the addition of 1% buffered sodium citrate and diacetate, without the outgrowth of *C. perfringens*.

Citric acid is sprayed on RTE products in edible and inedible casings, up to 10% solution, prior to slicing. Citric acid is also sprayed on the fibrous casings of RTE products (up to 3% solution) immediately before the casings are removed (USDA-FSIS, 2007).

Mixtures of organic acids have been found to be more effective in controlling *L. monocytogenes* on RTE meat products, than the use of individual organic acids (Palumbo and Williams, 1994; Samelis et al., 2002). Surface rinsing of pork and beef frankfurters with either a 2% acetic acid solution or a mixture of 1% acetic and 1% citric acids resulted in between 0.5- and 1-log reduction in *L. monocytogenes* on the frankfurters stored at 5°C (41°F) for up to 56 days (Palumbo and Williams, 1994).

Dipping frankfurters in either 2% acetic acid or a mixture of 1% acetic and 1% citric acids after frankfurters are inoculated with *L. monocytogenes* has been shown to be more effective in eliminating *L. monocytogenes* than if the frankfurters were dipped in the acid solutions before being inoculated with *L. monocytogenes* (Palumbo and Williams, 1994). The effect of dipping frankfurters in solutions of 1–2% acids was found to be bactericidal against *L. monocytogenes*; however, dipping frankfurters in 5% solutions of either lactic or acetic acids or a mixture of 2.5% acetic and 2.5% citric acids had both a bactericidal and bacteriostatic effect on *L. monocytogenes*, by not only reducing *L. monocytogenes* levels but also preventing the growth of *L. monocytogenes* over 90 days of storage at 5°C (41°F) (Palumbo and Williams, 1994).

A mixture of acetic, lactic, benzoic, and propionic acids has been shown to be effective in controlling *L. monocytogenes*, which applied to RTE frankfurters in combination with steam pasteurization (Murphy et al., 2006).

Fatty Acids

Fatty acids, such as octanoic acid, have been shown to have antimicrobial properties when applied to the surface of RTE meat products (Burnett et al., 2007). In this research, 1% octanoic acid solutions were acidified with either phosphoric acid (pH 2) or citric acid (pH 4), and applied to the surface of a variety of RTE meat products immediately before vacuum packaging. The packaged products were then passed through a shrink tunnel at 93°C for 2 and 7 seconds. Surface treatments of 1% octanoic acid resulted in more than a 2-log reduction in *L. monocytogenes* for all RTE products (cured ham, oven-roasted turkey breast, comminuted roast beef, and whole-muscle roast beef) tested, except for the oil-browned turkey breast (Burnett et al., 2007). This treatment has been commercialized as Octa-Gone by EcoLab (St. Paul, MN).

Acidifiers

Acidified Sodium Chlorite

Acidified sodium chlorite (ASC) is produced by acidifying sodium chlorite ($NaClO_2$) with a generally recognized as safe (GRAS) acid (Beverly et al., 2006) and it has been tested primarily for its efficacy as an antimicrobial agent for fresh or raw meat products.

ASC has been shown to be somewhat effective in controlling *L. monocytogenes* on cook-in bag turkey breasts (Luchansky et al., 2004). ASC has also be shown to reduce *L. monocytogenes* on cooked roast beef held at 4°C (39°F) for 28 days (Beverly et al., 2006).

Acidic Calcium Sulfate

Acidic calcium sulfate (ACS) is a GRAS ingredient that is approved for use in meat products. Dipping frankfurters in an ACS–lactic acid mixture was reported to significantly reduced *L. monocytogenes* (Keeton et al., 2002). This mixture also prevented any additional growth of *L. monocytogenes* for 12 weeks when stored at 4.5°C (40°F).

A mixture of ACS and lactic acid, when applied using the "sprayed lethality in container" (SLIC) system, only reduced *L. monocytogenes* 1–2 $\log_{10}$ CFU/ham within 24 hours on hams; however, this ACS–lactic acid treatment prevented further growth of *L. monocytogenes* for up to 60 days, when the hams were stored at 4°C (40°F) (Luchansky et al., 2005).

Chlorine Dioxide

Chlorine dioxide (at either 3 or 30 ppm concentrations) was not effective in reducing *L. monocytogenes* in "spent" brines, that had been used in chilling hot dogs and hams, most likely due to the organic matter in the used brine solutions (Valderrama et al., 2008).

Bacteriocinogenic Cultures

Organisms that naturally produce bacteriocins, to inhibit other bacteria in their environment, are described as being bacteriogenic. Nearly a fourth of the lactobacilli that were isolated from fermented sausages were found to be bacteriocinogenic (Garriga et al., 1994). Bacteriocinogenic starter cultures can be added to meat products before cooking, sprayed on the surface of the product, or applied to the inner surface of packaging films (Aymerich et al., 2006).

Pediococcus starter cultures have been found to reduce levels of *L. monocytogenes* in chicken summer sausage (Luchansky et al., 1992). Using *Lactobacillus sake*, as a starter culture, in making fermented sausage, has resulted in the production of bacteriocins, which inhibited *L. monocytogenes* (Tantillo et al., 2002). Adding *L. curvatus* and *Lactococcus lactis* susp. *lactis* with a commercial starter culture to a meat mixture, resulted in reduction in *L. monocytogenes* in the final fermented product (Benkerroum et al., 2005).

Bacteriocinogenic cultures of *Pediococcus acidilactici*, added to wieners and frankfurters, produced pediocin, and controlled the growth of *L. monocytogenes* in vacuum-packaged products (Berry et al., 1991; Degnan et al., 1992).

Bacteriocins

Bacteriocins are polypeptides that are produced by some bacteria, as a means of self-preservation, by inhibiting other bacteria in their environment. Bacteriocins, produced from lactic acid bacteria (LAB), should work well in food products as the producing organisms have GRAS status, they are not known to have any toxic effects, they are resistant to heat and prevent the growth of both Gram-positive spoilage and pathogenic organisms (Glvez et al., 2007). There are many types of bacteriocins, which inhibit *L. monocytogenes* and other Gram-positive organisms, including nisin, pediocin, enterocin, lacticin, etc. Bacteriocins could be added to RTE meat products by incorporating them in an active package, or by dipping or spraying them onto a product surface.

Nisin

Nisin is produced by *L. lactis* subsp. *lactis* and is the only bacteriocin currently approved as GRAS by the Food and Drug Administration (FDA), and has been approved for use in foods since 1969 (Delves-Broughton et al., 1996). Nisin is bactericidal (not bacteriostatic), in its effect on *L. monocytogenes*, but nisin may not be effective when used as the only antimicrobial agent, particularly for extended storage times (Jofre et al., 2008). Nisin has also been shown to be sporostatic, against the outgrowth of Clostridia spores, but needs to be active throughout the shelf life of the product to prevent the outgrowth of the spores (Joerger et al., 2000).

Adding nisin (800 AU/g; Niaplin) to the raw sausage batter was shown to significantly reduce *L. monocytogenes* counts in frankfurters over 60 days of storage at 3.5°C (38°F) (Hugas et al., 2002).

Surface application of nisin (250 μg/mL) to cooked pork tenderloin resulted in more than a 2-log CFU/g reduction in *L. monocytogenes*, with or without the use of modified atmosphere packaging (MAP) (Fang and Lin, 1994).

Nisin applied at 800 AU to cooked slices of a cured, vacuum-packaged, pork shoulder product, maintained the *L. monocytogenes* levels at the inoculated levels for 75 days, when stored at 1°C (34°F) (Jofre et al., 2008). In this same study, nisin was not effective in inhibiting *L. monocytogenes* when the sliced, pork products were stored at 6°C (43°F).

When nisin (Nisaplin) was sprayed onto the surface of peeled frankfurters, *L. monocytogenes* levels remained low (<3 MPN/g) over a 60-day storage time (Hugas et al., 2002).

The type of phosphate added to sausage products has been shown to affect the efficacy of nisin to inhibit LAB (Davies et al., 1999). Diphosphate, combined with nisin, was more effective against LAB than orthophosphate. In this same study, nisin was less effective in inhibiting LAB in sausage products of higher fat contents. However, it is not known what impact phosphate type or fat content has on inhibiting *L. monocytogenes* in sausage products.

When nisin was sprayed onto the surface of peeled frankfurters, *L. monocytogenes* was significantly reduced and this low level (<3 MPN/g) was maintained for up to 60 days (Hugas et al., 2002).

Nisin has also been added as a coating to the inside of sausage casings, with the intention that the antimicrobial effect would be transferred to the surface of the sausage during the cooking process. NoJax AL was a commercial example of this technology, but this product is currently not commercially available, due primarily to a lack of industry interest in this product (M.D. Nicholson, 2008, personal communication).

However, in other studies, casings containing 5,000 IU/g nisin were not found to inhibit *L. monocytogenes* on beef and turkey surfaces (Ming et al., 1997).

Nisin has been shown to be effective in controlling *L. monocytogenes* growth in RTE chicken when used in an edible zein coating, followed by a traditional vacuum package (Janes et al., 2002).

Vacuum-packaging barrier films, coated with a methylcellulose/hydroxypropyl methylcellulose-based solution, which contained

7,500 and 10,000 IU/mL of nisin reduced *L. monocytogenes* by more than 2 log CFU/package in hot dogs (Franklin et al., 2004).

Pediocin

Adding pediocin (800 AU/g; ALTA 2351) to raw sausage batter significantly reduced *L. monocytogenes* counts in frankfurters over 60 days of storage at 3.5°C (38°F) (Hugas et al., 2002).

Pediocin-coated (ALTA 2341) casings have also been shown to reduce *L. monocytogenes* on the surface of frankfurters and delayed the growth of *L. monocytogenes* during refrigerated storage (Chen et al., 2004). Pediocin-coated cellulose casings (9.3 $\mu g/cm^2$) and plastic bags ($7.75\ mg/cm^2 = 5\ AU/cm^2$) inhibited the growth of *L. monocytogenes* in beef, ham, and turkey breasts for 12 weeks, when stored at 4°C (40°F) (Ming et al., 1997). Pediocin applied to sliced RTE sausage reduced the level of *L. monocytogenes* by less than 1 log, and that level was maintained for up to 21 days (Mattila et al., 2003).

Other Bacteriocins and Cystibiotics

Enterocin A was found to be effective in inhibiting *L. inocua* in RTE ham, pork, chicken, pate, and sausage when the product was stored at 7°C (45°F) for 37 days, and in fermented sausage for 28 days (Aymerich et al., 2000a, 2000b). Adding 3,500 AU/g enterocin A to raw sausage batter (before cooking) was claimed to be more effective in reducing *L. monocytogenes* counts in frankfurters (over 60 days of storage at 3.5°C, or 38°F) than nisin (Hugas et al., 2002).

Enterocins A and B were shown to delay the growth of *L. monocytogenes* in vacuum-packaged cooked ham slices, when added to biodegradable films, made from alginate, zein, and polyvinyl alcohol. However, the enterocins were not able to effectively inhibit *L. monocytogenes* over long storage times (Marcos et al., 2007).

Other bacteriocins that have been shown to be effective in inhibiting *L. monocytogenes* growth in RTE meat products include Lactocin 705, Reuterin, and Sakacin (Doyle, 1999; Hugas et al., 1998; Krekel, 1997); however, these ingredients are not currently on the USDA-FSIS Table of Safe and Suitable Ingredients (USDA-FSIS, 2007).

There are limitations to the use of bacteriocins, particularly, if used without other antimicrobial agents. Most of the bacteriocins are not active against a broad range of Gram-positive pathogens, and are typically not effective against Gram-negative spoilage or pathogenic organisms. For this reason, the hurdle concept, mentioned earlier, is necessary, which would involve combining the use of bacteriocins with high-pressure processing (HPP), irradiation, etc. (Aymerich et al., 2006).

Antimycotic Agents

Antimycotic agents such as sorbate, benzoate, and propionate are used on meat products to prevent the growth of molds, during the long drying and distribution processes of these products. They have also been shown to inhibit Gram-positive organisms, such as Listeria, in meat products (Samelis et al., 2001); however, they are not currently approved for addition to a meat product formulation.

Potassium sorbate is the most common form of sorbates used in the food industry. It has been reported to be more effective than either benzoate or propionate in preserving food products; however, sorbates are currently approved only for use as an antimold ingredient for surface application (2.5% solution) to dry sausage products (Sofos and Busta, 1980).

L. monocytogenes did not grow on a cured beef and pork bologna when combinations of either 0.05% sodium benzoate and 0.05% potassium sorbate or sodium benzoate and sodium propionate were added to the formula (Glass et al., 2007). Adding these same combinations of ingredients to an uncured, turkey bologna inhibited, but did not prevent *L. monocytogenes* from growing, which indicated that these antimycotic agents were more effective when used in products that contained sodium nitrite (Glass et al., 2007).

L. monocytogenes growth was significantly inhibited in cured ham slices that contained 0.1% sodium benzoate, 0.2% sodium propionate, 0.3% sodium propionate, 0.3% potassium sorbate, 0.1% sodium propionate combined with 0.1% potassium sorbate, or 1.6% lactate combined with 0.1% diacetate, when stored at 4°C (39°F) for 12 weeks (Glass et al., 2007). These antimycotic ingredients were less effective at preventing *L. monocytogenes* growth when the sliced ham was stored at 7°C and 10°C (45°F and 50°F). In this study, consumer panels preferred the flavor of the propionate- and benzoate-treated ham over the lactate- or diacetate-treated ham.

Higher concentrations of antimycotic ingredients were needed to prevent *L. monocytogenes* growth in uncured turkey breast slices, compared to cured ham slices (Glass et al., 2007). More than 0.2% sodium propionate, combinations of more than 0.1% sodium proprionate and 0.1% potassium sorbate, or the combinations of 3.2% sodium lactate and 0.2% sodium diacetate were needed to inhibit *L. monocytogenes* growth to less than a 1-log increase in uncured turkey breast slices which were stored at 4°C (40°F) for 12 weeks.

Although sorbate was not effective in inhibiting clostridium in other foods, 0.1% potassium sorbate has been shown to reduce *C. perfringens* to undetectable levels and inhibited *C. botulinum* growth and toxin production in uncured cooked sausage that was temperature abused at 27°C (81°F) (Tompkin, 1978). The anticlostridium effect seems to be on spore germination and outgrowth (Sofos and Busta, 1980).

Sorbates were also proposed as a partial replacement of nitrite in cured meats, and a combination of 0.2% sorbate combined with 40–80 ppm nitrite was shown to extend the time for botulinum toxin production in meat products that were temperature abused (Ivey et al., 1978). In 1979, a USDA regulation was proposed to allow the use of 0.26% potassium sorbate in combination with 40 ppm nitrite in bacon; however, because of some allergic reactions that some sensory panelists experienced with consumption of this combination, this regulation was not issued (Sofos and Busta, 1980).

Ozone

Ozone is generated by passing oxygen through a high-voltage electrical field and this gas is claimed to be a more effective sanitizer than chlorine for some food surface applications, particularly in the presence of high organic matter (Cutter, 1998).

A gaseous environment of ozone (0.5–1.0 ppm) has been shown to be effective in inactivating *L. monocytogenes* on cured ham slices (Jhala et al., 2002). However, because of the potential hazards associated with exposing people to the gaseous ozone, the use of gaseous ozone is limited in the food industry. As an alternative, ozonated water is used as an effective decontaminating agent on food surfaces (Cutter, 1998).

Spices

Spices have been shown to have some antimicrobial properties in meat products. Clove oil was shown to significantly reduce *L. monocytogenes* in hot dogs of all fat levels (Singh et al., 2003). In the same study, thyme oil was not as effective as clove oil, but was shown to reduce *L. monocytogenes* in zero-fat hot dogs, but was less effective in low-fat (9% fat) hot dogs, and not effective in full-fat (26% fat) products.

At 0.5 and 1.0% levels in tryptic soy broth, cloves have been found to be bactericidal and oregano was bacteriostatic to *L. monocytogenes* at 4 and 24°C (40 and 75°F) (Ting et al., 1991). Sage was bactericidal at 4°C (40°F) and bacteriostatic at 24°C (75°F). However, when tested in sterile meat slurry at a 1% concentration, neither cloves nor oregano prevented *L. monocytogenes* growth.

Essential oil from mint was not found to prevent *L. monocytogenes* growth in pate at 4 and 10°C (40 and 50°F) (Tassou et al., 1995). Essential oils of spices (e.g., oregano and garlic) have been shown to be effective in inhibiting *L. monocytogenes* when added to edible films (Seydim and Sarikus, 2006).

A trend observed with most spice research shows that levels of spices needed to prevent *L. monocytogenes* in foods high in protein or fat are higher than would commonly be used in meat products (Ting and Deibel, 1991).

Smoke Treatments

Natural and liquid smoke treatments are known to have antimicrobial effects on cooked meat products. However, the residual effectiveness of preventing *L. monocytogenes* growth over the shelf-life of RTE meat products has been debatable. Earlier research had identified isoeugenol, a phenolic compound in liquid smoke products, as the active ingredient in liquid smoke products which was preventing the growth of *L. monocytogenes* in cooked meat products (Faith et al., 1992). However, more recent research found that separating the carbonyl compounds of liquid smoke from the phenols and acids, resulted in an extract which inhibited *L. monocytogenes* growth when sprayed onto RTE product surfaces immediately before packaging, but does not add additional smoke flavor (Gedela et al., 2007).

Liquid smoke, combined with 10% acetic acid, has been shown to inhibit *E. coli* O157:H7, *Salmonella*, *Listeria monocytogenes*, and *Streptococcus* on RTE meat products (Samelis et al., 2001). Liquid smoke, combined with steam, was effective in destroying *L. monocytogenes* and inhibiting the growth of *L. monocytogenes* in frankfurters (Murphy et al., 2005b).

Electrolyzed Oxidizing Water

Electrolyzed oxidizing (EO) water is generated by electrolysis of a saline solution, and a 7-log reduction in *L. monocytogenes* was reported when the pathogen was exposed to the water for 5 minutes (Venkitanarayanan et al., 1999). However, when sprayed on frankfurters and ham, acidic EO water showed some effect (not more than 1 $\log_{10}$ CFU/g) in inhibiting *L. monocytogenes* growth, but this level of inhibition is not sufficient to satisfy FSIS requirements as an antimicrobial agent to control *L. monocytogenes* growth in RTE meat products (Fabrizio and Cutter, 2005).

Both acidic and alkaline EO water have been shown to be effective in reducing *L. monocytogenes* biofilms (Ayebah et al., 2006).

Lauric Arginate

Lauric arginate, also known as ethyl-*N*-dodecanoyl-L-arginate hydrochloride, has been shown to be effective in reducing *L. monocytogenes* levels on the surface of RTE meat products, when added to the product formula (not to exceed 200 ppm of the finished product), sprayed onto RTE product prior to packaging, or sprayed inside the pouch at packaging. Lauric arginate has not been effective in controlling the growth of *L. monocytogenes* over the refrigerated shelf life of vacuum-packaged RTE meat products.

Luchansky et al. (2005) developed a spray lethality in container (SLIC^{TM}) process, which involved treating the purge within vacuumized, RTE meat product packages, rather than the product surface, with an antimicrobial, such as lauric arginate, to destroy pathogens on the surface of the products (e.g., *L. monocytogenes*). In a vacuum package, the purge is distributed evenly over the entire product surface, inside the package. A total of 2–8 mL of 5% lauric arginate, added to the package before inoculated hams was vacuum packaged, resulted in a reduction in *L. monocytogenes* from 3.3 to 6.5 $\log_{10}$/ham after 24 hours at 4°C (39°F). Hams treated with 10% solutions of lauric arginate resulted in 6.5 $\log_{10}$/ham after 24 hours at 4°C (39°F).

In shelf-life studies, Luchansky *et al.* (2005) found that whereas the lauric arginate treatment reduced *L. monocytogenes* after 24 hours, at 4°C (39°F), *L. monocytogenes* levels increased from 2.0 to 4.6 $\log_{10}$/ham when treated with 5% lauric arginate after 60 days. However, with hams inoculated with 7 $\log_{10}$/ham *L. monocytogenes*, lauric arginate treatments resulted in significantly lower levels of *L. monocytogenes* after 60 days at 4°C (39°F), than the untreated control hams, and in all cases, the *L. monocytogenes* levels were lower than the initial inoculation levels (Luchansky et al., 2005).

Using a lower inoculation level (ca.3 log/ham) of *L. monocytogenes*, it was found that a 5% lauric arginate treatment reduced *L. monocytogenes* levels to below the limit of detection after 24 hours at 4°C (39°F), and the highest level of addition (8 mL) prevented further growth of *L. monocytogenes* for an additional 60 days (Luchansky et al., 2005).

Combining the listeriocidal effects of lauric arginate with the listeriostatic effects of lactate and diacetate has been shown to be most effective in preventing the growth of *L. monocytogenes* over 90 days of refrigerated storage, and this combination would allow RTE products to be made under Alternative 1 (S. Veasey, 2008, personal communication).

Bacteriophages

A mixture of bacteriophages, specific for *L. monocytogenes*, can be applied as a spray at a level not to exceed 1 mL per 500 cm^2 of product surface area (21 CFR 172.785). The addition of 7 or 9 log_{10} of P100 bacteriophage reduced initial levels of *L. monocytogenes* by 1 log, and prevented the growth of *L. monocytogenes* for up to 21 days of storage at 4°C (40°F); however, the growth of *L. monocytogenes* increased at 28 days at the above storage conditions (Call et al., 2008). When bacteriophage was combined with varying levels of potassium lactate and sodium diacetate (0.68%:0.097%, 1.02%:0.145%, and 1.36%:0.19%), the growth of *L. monocytogenes* was essentially unchanged over a 45-day storage period at either 4 or 10°C (40 or 50°F) (Call et al., 2008).

Antimicrobial Processes

Ohmic Heating

Ohmic heating involves passing electrical currents through meat products, which heats the product internally because of the electrical resistance of the product (Vicente et al., 2006). Early attempts found ohmic heating of meat to be problematic. One potential application was an indirect thawing process for frozen meat using a liquid-contact procedure (Wang et al., 2002).

Ohmic heating of sausage and ham products was done in a batch process (Piette and Brodeur, 2001). A 1-kg ham was cooked in less than 2 minutes; however, level of lethality was not sufficient to guarantee a safe RTE product.

Active Packaging

Active packaging involves incorporating antimicrobial agents into packaging materials during film extrusion, coating the surface layer of the packaging film, which touches the product, or into a sachet or absorbent pad, which would be placed inside the package with the product. The antimicrobial agent is gradually released from the film, sachet, or absorbent pad, during the shelf life of the product, inhibiting *L. monocytogenes* which potentially could have contaminated the product between cooking and packaging.

Since the greatest risk to RTE meat products is surface contamination by *L. monocytogenes*, use of antimicrobials in some form of active packaging would appear to reduce the quantity of antimicrobial agent needed, compared to adding antimicrobials to the product formulation (Appendini and Hotchkiss, 2002). Gradually releasing the antimicrobial agents from the film might provide more antimicrobial activity than dipping meat products in antimicrobial agents prior to packaging. Some antimicrobial agents may lose effectiveness or be diluted by the product purge in the case of dipped or sprayed products (Appendini and Hotchkiss, 2002). Organic acids, and bacteriocins (particularly nisin), have been used in active packaging applications (Aymerich et al., 2008).

Hot dogs, vacuum packaged in films coated with a methylcellulose/hydroxypropyl methylcellulose-based solution containing 10,000 and 7,500 IU/mL of nisin, reduced *L. monocytogenes* by more than 2 log CFU/package and maintained this level for 60 days of refrigerated storage (Franklin et al., 2004).

L. innocua was reduced by 2 log CFU/g in cooked ham that was packaged in a polyethylene/polyamide package, with an absorbent cellulose pad, both of which were soaked in Nispalin® solutions (Scannell *et al.*, 2000). Pediocin-coated vacuum barrier bags reduced *L. monocytogenes* numbers over 12 weeks of storage at 4°C (40°F) (Ming et al., 1997).

The growth of *L. monocytogenes* was prevented on slices of beef bologna for up to 28 days at 4°C (40°F), when the bologna slices were wrapped in polyvinylidene chloride films containing sorbic acid (Limjaroen et al., 2005).

Chitosan-based antimicrobial packaging films have been shown to have antipathogen properties. Chitosan could be used as an edible film, and its application in combination with acetic and propionic acids has been studied; however, the release of these organic acids from the film has been shown to be slower at refrigerated temperatures (Ouattara et al., 2000).

MAP is also used for RTE meat products; however, the surface contact of antimicrobial agents would not apply to MAP meat products. McMullen and Stiles (1996) proposed the use of bacteriogenic cultures to produce bacteriocins to prevent pathogen growth/survival in MAP.

Packaging materials can be treated to have antimicrobial activity using radiation methods that include use of radioactive packaging films, ultraviolet (UV) light, or laser-excited materials. However, these methods have not been approved by FDA for use in meat products (Rourke, 2001).

The cost of using many of the antimicrobial agents in packaging films has limited the application of this technology in the USA (L.E. Cook, 2008, personal communication). Perhaps the most used antimicrobial additive to packaging films is the silver-substituted zeolites (Appendini and Hotchkiss, 2002). This technology has been used mostly in Japan, where the concern for food safety takes priority, regardless of product cost. The only antimicrobial additive that is approved by FDA for direct food contact is Zeomic (Appendini and Hotchkiss, 2002). Commercial examples of silver-substituted zeolites include Zeomic, Apacider, AgIon, Bactekiller, and Novaron (Brody et al., 2001).

Edible Films

Edible films made of hydroxyl propyl methyl cellulose (HPMC), to which bacteriocins were added to the mixture, and applied to cooked meat products, have been shown to be effective in inhibiting *L. monocytogenes* (Jacobsen et al., 2002).

Wheat gluten films, containing 5% Nisaplin®, were found to reduce *L. monocytogenes*, after 5 weeks on postpackage pasteurized bologna slices. This resulted in a 2.5-log CFU/g reduction in *L. monocytogenes* after 8 weeks (McCormick, 2001).

Nonthermal Postpackaging Treatments

Most *L. monocytogenes* problems have originated from environmental contamination of RTE meat products after the products have received an adequate lethality treatment, when the RTE products are exposed to the environment during the packaging process. The solutions to postlethality contamination by pathogens include aseptic processing or postpackaging reheating.

A postpackaging lethality treatment offers the advantage of destroying the surface of *L. monocytogenes* contamination that may have occurred after the original cooking process. There are a number of effective post-packaging reheating methods available, but none of these methods should be used as a substitute for good sanitation and handling procedures. Post-packaging reheating procedures should be used as additional pathogen intervention strategies to providing the safest RTE products possible to consumers.

Ultraviolet Light

UV light does not penetrate the surface of solid foods, such as meat; however, contamination by *L. monocytogenes* on cooked meat products would be primarily on the surface of these products. Shorter wavelengths (254 nm) have been shown to be more effective against *L. monocytogenes* than longer wavelengths (365 nm), and higher doses of UV irradiation (e.g., 550 $\mu W/cm^2$) inactivated *L. monocytogenes* more quickly than lower doses (100 $\mu W/cm^2$) (Yousef and Marth, 1988). UV irradiation (500 $\mu W/cm^2$, up to 3 minutes) was not effective in reducing *L. monocytogenes* on the surface of cooked chicken, with or without skin (Kim et al., 2002).

White light flashes, lasting for only a few hundred microseconds, generated at a wavelength of 200 nm in the UV spectrum to about 1 mm in the near-infrared range has also been found to be effective in reducing pathogens on meat products (Dunn et al., 1995). This process, commercially known as PureBright®, has been reported to cause a 1- to 3-$\log_{10}$ reduction in *Salmonella* spp. and *L. innocua* on chicken wings and frankfurters, respectively (Dunn et al., 1995). *L. monocytogenes* has been shown to be the most resistant of food-related pathogens to UV light (Rowan et al., 1999).

High-Pressure Processing

HPP of meat products involves applying water pressure (100–900 MPa) to the products. HPP is currently applied as either a batch or semicontinuous process. A pasteurization treatment requires from 300 to 600 MPa of pressure (Aymerich et al., 2008). Inactivation of vegetative pathogens with HPP is increased at higher (50–70°C, or 122–158°F) or lower (below 0°C, or 32°F) than room temperatures (Cheftel and Culioli, 1997).

For a sterilization process, over 600 MPa of pressure would be required, in combination with temperatures of 80–100°C (176–212°F), in order to inactivate spores (Cheftel, 1995; Cheftel and Culioli, 1997). HPP is known to be less effective against spores, compared to vegetative cells; however, the use of HPP has been shown to result in a 12D for *C. botulinum*, with the final product of higher quality and nutrient retention, compared to products retorted by conventional methods (Master et al., 2004). Use of HPP, combined with slightly elevated product temperatures has also been shown to be effective against *C. botulinum* type E spores (Rhodehamel and Beckwith, 1999). HPP is nearly an instantaneous treatment and the time required for this process is independent of the product size (Cheftel and Culioli, 1997).

HPP has been shown to effectively inhibit *L. monocytogenes* from growing in RTE meat products; however, Gram-positive organisms, such as *L. monocytogenes*, are known to be less sensitive to inactivation by irradiation than Gram-negative organisms (Suzuki et al., 2006). HPP applied to packaged RTE meat products at a pressure of 87,000 psi caused a 5- to 7-$\log_{10}$ reduction in both *Salmonella* and *L. monocytogenes* on sliced roast beef and turkey stored at 2°C (35°F), over 27 days (Rourke, 2001).

HPP is more effective in inactivating *L. monocytogenes* at colder product temperatures. HPP is often studied in combination with the use of other antimicrobials to maximize the safety of RTE meat products (Jofre et al., 2008).

Most studies show that HPP of packaged, RTE meat products is not detrimental to product flavor, but has been shown to increase free water or purge in the package (Pietrazak et al., 2007). More specifically, fresh meat color appears to lighten as a result of HPP. However, cured RTE meat color is not affected by HPP, which would make sliced, cured RTE products good candidates for HPP (Cheftel and Culioli, 1997).

One disadvantage to using HPP in RTE meat products is the limited production capacity, due to the batch nature of the process, as well as the time required for pressurization and discharge (Aymerich et al., 2008). For example, HPP equipment can handle 25–30 hams per basket (Suzuki et al., 2006). USDA-FSIS does not require that HPP be present on labels on HPP-treated meat products. HPP may also encourage oxidation of unsaturated fatty acids in meat products (Cheftel and Culioli, 1997).

Irradiation

Irradiation of meat products is a nonthermal process, which can be applied to products after packaging, to destroy potential *L. monocytogenes* contamination occurring between cooking and packaging and can be applied to either frozen or nonfrozen meat products (Ahn et al., 2006).

Irradiation involves exposing the product to ionizing radiation to destroy pathogens. Ionizing radiation has been shown to effectively destroy *L. monocytogenes*, and its lethality effects have been further enhanced when combined with various antimicrobial compounds (Ouattara and Mafu, 2000) and antimicrobial processes (Ahn et al., 2006).

The composition of meat products has been shown to affect the microbial decontamination of irradiation (Ahn et al., 2006). Increasing protein content appears to protect pathogens from the lethal effects of irradiation. However, fat content of meat products does not appear to affect the destruction of pathogens.

Meat product temperature affects the destruction of pathogens by irradiation. Product temperatures, below the freezing point, have a protective affect against the effects of irradiation (Ahn et al., 2006). The type of packaging applied to meat products has resulted in inconsistent effect of irradiation in destroying *L. monocytogenes* (Ahn et al., 2006).

Cured deli meats irradiated to 1.2–1.6 kGy have been found to result in acceptable color, flavor, and texture after 6 months of refrigerated storage; however, uncured irradiated roast beef and turkey products were reported to be less desirable in flavor (Rourke, 2001). The level of off-flavor development in uncured irradiated RTE meat products has been found to vary with consumers tested, but most could detect a flavor difference in the irradiated versus nonirradiated uncured products (Rourke, 2001). However, irradiation is currently not approved for use on products that are processed (contain salt, nitrite, etc.) and/or RTE.

Thermal Postpackaging Treatments

Postpackaging Pasteurization

The most common approach to postpackaging pasteurization in the USA involves applying hot water or steam to RTE meat packages, in an attempt to eliminate *L. monocytogenes*. Murphy and Berrang (2002) claimed that at 88°C (190°F), there was no difference in lethality of *L. monocytogenes* on chicken breast strips or treatment time required, when either steam or hot water were used in postpackaging pasteurization. In addition to the elimination of *L. monocytogenes*, some companies have experienced increased shelf life of postpackaging pasteurized products by 25–33% (Rourke, 2001).

A guideline, based on research to date, for postpackaging pasteurization of RTE meat products (that weigh 20 lb/9 kg or less) using 96°C (205°F) water, for 10 minutes, should result in a 2- to 4-log reduction in *L. monocytogenes* (Houben and Eckenhausen, 2006). Cooking boneless chicken breasts under moist conditions resulted (Harrison and Carpenter, 1989) in a 1.86-$\log_{10}$ reduction in *L. monocytogenes* when cooked to 71.1°C (160°F) internal temperature, a 4-$\log_{10}$ reduction in *L. monocytogenes* at 73.9°C (165°F), a 5.54-$\log_{10}$ reduction in *L. monocytogenes* at 76.7°C (170°F), and a 5.42-$\log_{10}$ reduction in *L. monocytogenes* at 82.2°C (180°F).

Murphy et al. (2003a) predicted the treatment time required to obtain adequate *L. monocytogenes* lethality for vacuum-packaged chicken breasts of different thicknesses. Lethality of postpackaging reheating may not be as high as expected based only on product surface temperatures (Houben and Eckenhausen, 2006), and this might be due to the surface irregularities of RTE meat products.

Packaging films used for postpackaging pasteurization must be able to resist water or steam temperatures of 88–96°C (190–205°F) for the time required to obtain adequate lethality of *L. monocytogenes* (Rourke, 2001).

Whereas the surface temperature of RTE meat products is normally the temperature of concern when developing postpackaging reheating procedures for RTE meat products, some protocols, particularly with small diameter RTE meat products, processors may need to reheat these products internally to lethality temperatures, so product quality must be considered when applying this technology to RTE meat products. With some products, melted fat may be observed to accumulate on the surface of the products, which would reduce consumer acceptability. However, such high surface temperatures may not be needed when postpackaging reheating is applied to RTE meat products, which contain antimicrobial ingredients mentioned earlier.

Muriana et al. (2002) found that additional cold purge from the center of RTE meat products, during the postpackaging pasteurization, reduced the effectiveness of the process. However, Houben and Eckenhausen (2006) contend that this conclusion contradicts what is typically understood from sterilization technology, because a moist environment destroys more spores than a dry environment.

Steam pasteurization applied to RTE meat product surfaces during vacuum packaging has been developed so that postlethality treatments can be applied at a production line speeds (Kozempel et al., 2000; Murphy et al., 2005a). Applying an organic acid solution of 2% acetic, 1% lactic, 0.1% propionic, and 0.1% benzoic acids, with steam surface pasteurization (114°C for 1.5 seconds), during vacuum packaging resulted in a 3-log reduction in *L. monocytogenes* (Murphy et al., 2006). This combination of organic acid treatment and steam pasteurization also inhibited the growth of *L. monocytogenes* in vacuum-packaged frankfurters for 19 and 14 weeks, when stored at 4 and 7°C (40 and 45°F), respectively.

Muriana et al. (2004) reported that combining radiant heating (399°C/750°F) on exposed RTE turkey products immediately prior to packaging, followed by postpackaging pasteurization while the product surface was still warm (27°C, or 81°F), resulted in a more effective lethality of *L. monocytogenes* and reduced purge loss from the product, compared to postpackaging pasteurization applied to cold products. Authors speculated that the effectiveness of the prepackaging pasteurization may be due to the effect of vacuum packaging potentially pulling *L. monocytogenes* contamination from product surfaces into cracks and crevices.

Murphy et al. (2003b) showed that the *L. monocytogenes* lethality temperature needs to be achieved as deep as 15 mm below the surface to allow for typical surface roughness.

It is important to use the proper *D* value when calculating the lethality of a postpackaging pasteurization treatment. The *D* values that were calculated during postpackaging pasteurization of fermented sausages were claimed to be more likely a result of the water temperatures used, rather than the endpoint surface temperature that was targeted for the product (Roering et al., 1998).

Postpackaging pasteurization of bologna resulted in a 3.5- to 4.2-log reduction in *L. monocytogenes* at the time of the heat treatment and a significant reduction in *L. monocytogenes* after 12 weeks of storage at 4°C (39°F) (Mangalassary et al., 2008). Combining postpackaging pasteurization with nisin–lysozyme treatments reduced *L. monocytogenes* to below detectable levels after 2–3 weeks.

The results from different postpackaging pasteurization studies are difficult to compare (Houben and Eckenhausen, 2006), because of the different methods used (i.e., cooking medium temperatures, product sizes, packaging film types and thicknesses, and varying product compositions) in each study and the varying details that were provided in the published literature resulting from each study.

Summary

Consumers are demanding safer meat products with longer shelf lives, as well as more naturally produced products, made without chemical preservatives. Biopreservation (e.g., use of bacteriogenic cultures and bacteriocins), postpackaging pasteurization, and lower product storage temperatures may be the solution to meeting consumer food safety demands (Aymerich et al., 2006).

References

Ahmed, O., White, T., and Draughon, F.A. 2008. HACCP validation for use of lactic acid on bologna, ham and red hot ready-to-eat meat and poultry products. In: *Abstracts of the Annual Meeting of the International Association for Food Protection (P2-43)*, p. 91.

Ahn, D.U., Lee, E.J., and Mendonca, A. 2006. Meat decontamination by irradiation. In: Nollet, L.M.L., and Toldra, F. (eds), *Advanced Technologies for Meat Processing*. CRC–Taylor & Francis Group, Abingdon, Oxfordshire, UK.

Appendini, P., and Hotchkiss, J.H. 2002. Review of antimicrobial food packaging. *Innovative Food Science and Emerging Technologies* 3:113–126.

Ayebah, B., Hung, Y.-C., Kim, C., and Frank, J.F. 2006. Efficacy of electrolyzed water in the inactivation of planktonic and biofilm *Listeria monocytogenes* in the presence of organic matter. *Journal of Food Protection* 69(9):2143–2150.

Aymerich, T., Artigas, M.G., Garriga, M., Monfort, J.M., and Hugas, M. 2000a. Effect of sausage ingredients and additives on the production of enterocin A and B by *Enterococcus faecium* CTC 492. Optimization of *in vitro* production and antilisterial effect in dry fermented sausages. *Journal of Applied Microbiology* 88:686–694.

Aymerich, T., Garriga, M., Jofre, A., Martin, B., and Monfort, J.M. 2006. The use of bacteriocins against meat-borne pathogens. In: Nollet, L.M.L., and Foldra, F. (eds), Advanced Technologies for Meat Processing. CRC–Taylor & Francis Group, Abingdon, Oxfordshire, UK.

Aymerich, T., Garriga, M., Ylla, J., Vallier, J., Monfort, J.M., and Hugas, M. 2000b. Application of enterocins as biopreservatives against *Listeria inocua* in meat products. *Journal of Food Protection* 63:721–726.

Aymerich, T., Picouet, P.A., and Monfort, J.M. 2008. Decontamination technologies for meat products. *Meat Science* 78(1–2):114–129.

Barmpalia, I.M., Koutsoumanis, K.P., Geornaras, I., Belk, K.E., Scanga, J.A., Kendall, P.A., Smith, G.D., and Sofos, J.N. 2005. Effect of antimicrobials as ingredients of pork bologna for *Listeria monocytogenes* control during storage at 4 or 10°C. *Food Microbiology* 22:205–211.

Benkerroum, N., Daoudi, A., Hamraoui, T., Ghalfi, H., Thiry, C., Duroy, M., Evrart, P., Roblain, D., and Thonart, P. 2005. Lyophilized preparations of bacteriocinogenic *Lactobacillus curvatus* and *Lactococcus lactis* susp. *lactis* as potential protective adjuncts to control *Listeria monocytogenes* in dry-fermented sausages. *Journal of Applied Microbiology* 98:56–63.

Berry, E.D., Hutkins, R.W., and Mandigo, R.W. 1991. The use of bacteriocin-producing *Pediococcus acidilactici* to control post-processing *Listeria monocytogenes* contamination of frankfurters. *Journal of Food Protection* 54:681–686.

Beverly, R.L., Janes, M.E., and Oliver, G. 2006. Research note: Acidified sodium chlorite treatment for inhibition of lmono growth on the surface of cooked roast beef. *Journal of Food Protection* 69(2):432–435.

Brewer, M.S., McKeith, F., Martin, S.E., Dallmier, A.W., and Wu, S.Y. 1992. Some effects of sodium lactate on shelf-life, sensory, and physical characteristics of vacuum-packaged beef bologna. *Journal of Food Quality* 15(5):369–382.

Brody, A., Strupinsky, E., and Kline, L. 2001. *Active Packaging for Food Applications*. Technomic Publishing Company, Lancaster, PA.

Buchanan, R.L., Stahl, H., and Whiting, R.C. 1989. Effects and interactions of temperature, pH, atmosphere, sodium chloride, and sodium nitrite on the growth of *Listeria monocytogenes*. *Journal of Food Protection* 52(12):844–851.

Burnett, S.L., Chopskie, J.H., Podtburg, T.C., Gutzmann, T.A., Gilbreth, S.E., and Bodnaruk, P.W. 2007. Use of octanoic acid as a postlethality treatment to reduce *Listeria monocytogenes* on ready-to-eat meat and poultry products. *Journal of Food Protection* 70(2):392–398.

Call, J.E., Porto-Fett, A.C.S., and Luchansky, J.B. 2008. Effectiveness of bacteriophage to control the outgrowth of *Listeria monocytogenes* on the surface of frankfurters. In: Poster at International Association of Food Protection Annual Conference, Columbus, OH.

Cheftel, J.C. 1995. Review: High-pressure, microbial inactivation and food preservation. *Food Science and Technology International* 1:75–90.

Cheftel, J.C., and Culioli, J. 1997. Effects of high pressure on meat: A review. *Meat Science* 46(3):221–236.

Chen, N., Sebranek, J., Dickson, J., and Mendonca, A. 2004. Combining pediocin with post-packaging thermal pasteurization for control of *Listeria monocytogenes* on frankfurters. *Journal of Food Protection* 67(9):1855–1865.

Chen, N., and Shelef, L.A. 1992. Relationship between water activity, lactate and growth of *Listeria monocytogenes* in a meat model system. *Journal of Food Protection* 55:574–578.

Cutter, C.N. 1998. New intervention technologies. *Proceedings of the 51st Annual Reciprocal Meat Conference* 51:133–140.

Davies, E.A., Milne, C.F., Bevis, H.E., Potter, R.W., Harris, J.M., Williams, G.C., Thomas, L.V., and Delves-Broughton, J. 1999. Effective use of nisin to control lactic acid bacterial spoilage in vacuum-packed bologna-style sausage. *Journal of Food Protection* 62:1004–1010.

Degnan, A.J., Yousef, A.E., and Luchansky, J.B. 1992. Use of *Pediococcus acidilactici* to control *Listeria monocytogenes* in temperature-abused vacuum-packaged wieners. *Journal of Food Protection* 55:98–103.

Delves-Broughton, J., Blackburn, P., Evans, R.J., and Hugenholtz, J. 1996. Applications of the bacteriocins, nisin. *Antonie van Leeuwenhoek* 69:193–202.

Doyle, M.E. 1999. *Use of Other Preservatives to Control Listeria in Meat. Food Research Institute Briefings: Control of Listeria in Meat.* University of Wisconsin.

Dunn, J., Ott, T., and Clark, W. 1995. Pulsed-light treatment of food and packaging. *Food Technology* 49:95–98.

Duxbury, D.D. 1988. Natural sodium lactate extends shelf-life of whole and ground meats. *Food Processing* 49:91.

Evans, L.L., Pagach, D.A., Belk, K.E., and Miller, R.K. 1991. Sodium lactate treated roast beef to prolong storage and enhance palatability [Abstract]. *Proceedings of the IFT Annual Meeting* 76:142.

Fabrizio, K.A., and Cutter, C.N. 2005. Application of electrolyzed oxidizing water to reduce *Listeria monocytogenes* on ready-to-eat meats. *Meat Science* 71(2):327–333.

Faith, N.G., Yousef, A.E., and Luchansky, J.B. 1992. Inhibition of Listeria mono by liquid smoke and isoeugenol, a phenolic component found in smoke. *Journal of Food Safety* 12(4):303–314.

Fang, T.J., and Lin, C.C. 1994. Growth of *L. monocytogenes* and *Pseudomonas fragi* in cooked pork in a modified atmosphere packaging/nisin combination system. *Journal of Food Protection* 57:479–485.

Farber, J.M. 1989. Thermal resistance of *Listeria monocytogenes* in foods. *International Journal of Food Microbiology* 8:285–291.

Franklin, N.B., Cooksey, K.D., and Getty, K.J.K. 2004. Inhibition of *Listeria monocytogenes* on the surface of individually packaged hot dogs with packaging film coating containing nisin. *Journal of Food Protection* 67:480–485.

Gálvez, A., Abriouel, H., López, R.L., and Omar, N.B. 2007. Bacteriocin-based strategies for food biopreservation. *International Journal of Food Microbiology* 120:51–70.

Garriga, M., Hugas, M., Aymerich, T., and Monfort, J.M. 1994. Bacteriocinogenic activity of lactobacilli from fermented sausages. *Journal of Applied Bacteriology* 75:142–148.

Gedela, S., Escoubas, J.R., and Muriana, P.M. 2007. Effect of inhibitory liquid smoke fractions on *L. monocytogenes* during long-term storage of frankfurters. *Journal of Food Protection.* 70(2):386–391.

Geornaras, I., Belk, K.E., Scanga, J.A., Kendall, P.A., Smith, G.D., and Sofos, J.N. 2005. Postprocessing antimicrobial treatments to control *Listeria monocytogenes* in commercial vacuum-packaged bologna and ham stored at 10°C. *Journal of Food Protection* 68:991–998.

Geornaras, I., Skandamis, P.N., Belk, K.E., Scanga, J.A., Kendall, P.A., Smith, G.D., and Sofos, J.N. 2006. Postprocessing application of chemical solutions for control of *Listeria monocytogenes*, cultured under different conditions, on commercial smoked sausage formulated with and without potassium lactate-sodium diacetate. *Food Microbiology* 23:762–771.

Glass, K.A., Granberg, D.A., Smith, A.L., McNamara, A.M., Hardin, M., Mattias, J., Ladwig, K., and Johnson, E.A. 2002. Inhibition of *Listeria monocytogenes* by sodium diacetate and sodium lactate on wieners and cooked bratwurst. *Journal of Food Protection* 65(1):11–123.

Glass, K.A., and McDonnell, L.M. 2008. Minimum nitrite level required to inhibit *Listeria monocytogenes* on ready-to-eat turkey prepared with lactate and diacetate. In: *Abstracts of the Annual Meeting of the International Association for Food Protection (P2-38)*, p. 89.

Glass, K.A., McDonnell, L.M., Rassel, R.C., and Zierke, K.L. 2007. Controlling *Listeria monocytogenes* on sliced ham and turkey products using benzoate, propionate, and sorbate. *Journal of Food Protection* 70(10): 2306–2312.

Glass, K.A., Preston, D., and Veesenmeyer, J. 2007. Inhibition of *Listeria monocytogenes* in turkey and pork-beef bologna by combinations of sorbate, benzoate, and propionate. *Journal of Food Protection* 70(1): 214–217.

Harrison, M.A., and Carpenter, S.L. 1989. Survival of large populations of *Listeria monocytogenes* on chicken breasts processed using moist heat. *Journal of Food Protection* 52(6):376–378.

Hofing, D., Brooks, R., Hull, R., and McClintock, R. 2008. Efficacy of four antimicrobial ingredients to inhibit the growth of *Listeria monocytogenes* in roast beef. In: *IAFP 2008 Technical Abstracts, T2-05*.

Houben, J.H., and Eckenhausen, F. 2006. Surface pasteurization of vacuum-sealed precooked ready-to-eat meat products. *Journal of Food Protection* 69(2):459–468.

Hugas, M., Garriga, M., Aymerich, M.T., and Monfort, J.M. 2002. Bacterial cultures and metabolites for the enhancement of safety and quality in meat products. In: Toldrà, F. (ed), *Research Advances in the Quality of Meat Products*. Research Signpost, Trivandrum, India, pp. 225–247.

Hugas, M., Pages, F., Garriga, M., and Monfort, J.M. 1998. Application of the bactericinogenic *Lactobacillus sakei* CTC 494 to prevent growth of Listeria in fresh and cooked meat products packed with different atmospheres. *Food Microbiology* 15:639–650.

Ivey, F.J., and Robach, M.C. 1978. Effect of sorbatic acid on *Clostridium botulinum* outgrowth and toxin production in canned comminuted pork. *Journal of Food Science* 43:1782.

Ivey, F.J., Shaver, K.J., Christiansen, L.N., and Tompkin, R.B. 1978. Effect of potassium sorbate on toxinogenesis of *Clostridium botulinum* in bacon. *Journal of Food Protection* 41:621–625.

Jacobsen, T., Koch, A.G., Graveson, A., and Knochel, S. 2002. Biocontrol of class IIa bacteriocin sensitive and resistant *Listeria monocytogene* on sliced meat products. In: Proceedings of the 18th International ICFMH Symposium Food Micro, Lillehammer, Norway, pp. 18–23.

Janes, M.E., Kooshesh, S.N., and Johnson, M.G. 2002. Control of *Listeria monocytogenes* on the surface of refrigerated, ready-to-eat chicken coated with edible zein films containing nisin and calcium proprionate. *Journal of Food Science* 67(7):2754–2757.

Jhala, R., Muthukumarappan, K., Julson, J.L., Dave, R.I., and Mahapatra, A.K. 2002. Synergistic effect of ozone and Microgard® 300 for controlling *Listeria monocytogenes z*in ready-to-eat cooked and cured ham. In: *ASABE Annual International Meeting/CIGR XVth World Congress*, Chicago, IL, July 28–31.

Joerger, R.D., Hoover, D.G., Barefoot, S.F., Harmon, K.M., Grinstead, D.A., and Nettles-Cutter, C.G. 2000. Bacteriocins. In: Lederberg, J. (ed), *Encyclopedia of Microbiology*, Vol. 1, 2nd edn. Academic Press, San Diego, pp. 383–397.

Jofre, A., Garriga, M., and Aymerich, T. 2008. Inhibition of Salmonella sp., *Listeria monocytogenes* and *Staphylococcus aureus* in cooked ham by combining antimicrobials, high hydrostatic pressure and refrigeration. *Meat Science* 89(1–2):53–59.

Keeton, J.T., Acuff, G.R., Nuñez de Gonzalez, M.T., Ringer, L.J., and Lucia, L.M. 2002. Antimicrobial effects of surface treatments and ingredients on cured RTE meat products. Final Report to the American Meat Institute Foundation, Washington, DC.

Kim, T., Silva, J.L., and Chen, T.C. 2002. Effects of UV irradiation of selected pathogens in peptone water and on stainless steel and chicken meat. *Journal of Food Protection* 65(7):1142–1145.

Kozempel, M., Goldberg, N., Radewonuk, E.R., and Scullen, O.J. 2000. Rapid hot dog surface pasteurization using cycles of vacuum and steam to kill *Listeria innocua. Journal of Food Protection* 63:457–461.

Kröekel, L. 1997. Use of sakacin P and *Lacobacillus sake* LB674 to suppress growth of *Listeria monocytogenes* in vacuum-packaged sliced bologna-type sausage. In: *Proceedings of the LACTIC97*, Villers-Bocage, France.

Leistner, L. 2000. Basic aspects of food preservation by hurdle technology. *International Journal of Food Microbiology* 55:181–186.

Limjaroen, P., Ryser, E., Lockhart, H., and Hart, B. 2005. Inactivation of *Listeria monocytogenes* on beef bologna and cheddar cheese using polyvinylidene chloride films containing sorbic acid. *Journal of Food Science* 70(5):267–271.

Luchansky, J.B., Call, J.E., Hristova, B., Rumery, L., Yoder, L., and Oser, A. 2005. Viability of *Listeria monocytogenes* on commercially-prepared hams surface treated with acidic calcium sulfate and lauric arginate and stored at 4°C. *Meat Science* 71:92–99.

Luchansky, J.B., Cocoma, G.J., and Call, J.E. 2004. Hot water post-processing pasteurization of cook-in-bag turkey breast treated with potassium lactate and sodium diacetate (Opti-Form) and acidified sodium chlorite for control of *Listeria monocytogenes* during storage at 4°C. In: *Abstracts of the Annual Meeting of the International Association for Food Protection (P001)*, p. 54.

Luchansky, J.B., Glass, K.A., Hasono, K.D., Degnan, A.J., Faith, N.G., Cauvin, B., Baccus-Taylor, G., Arihara, K., Bater, B., Maurer, A.J., and Cassens, R.G. 1992. Genomic analysis of *Pediococcus* starter cultures used to control *Listeria monocytogenes* in turkey summer sausage. *Applied and Environmental Microbiology* 58:3053–3059.

Maas, M.R., Glass, K.L., and Doyle, M.P. 1989. Sodium lactate delays toxin production by *Clostridium botulinum* in cook-in-bag turkey products. *Applied Environmental Microbiology* 55:2226.

Maca, J.V., Miller, R.K., Binger, M.E., Lucia, L.M., and Acuff, G.R. 1999. Sodium lactate and storage temperature effects on shelf life of vacuum packaged beef top rounds. *Meat Science* 53:23–29.

Mackey, B.M., Pritchet, C., Norris, A., and Mead, G.C. 1990. Heat resistance of *Listeria*: Strain differences and effect of meat type and curing salts. *Letters in Applied Microbiology* 10:251–255.

Mangalassary, S., Han, I., Rieck, J., Acton, J., and Dawson, P. 2008. Effect of combining nisin and/or lysozyme with in-package pasteurization for control of Listeria monocytogenes in ready-to-eat turkey bologna during refrigerated storage. *Food Microbiology* 25(7):866–870.

Marcos, B., Aymerich, T., Monfort, J.M., and Garriga, M. 2007. Use of antimicrobial biodegradable packaging to control *Listeria monocytogenes* during storage of cooked ham. *International Journal of Food Microbiology* 120:152–158.

Master, A.M., Krebbers, B., Van Den Berg, R.W., and Bartels, P.V. 2004. Advantages of high pressure sterilization on quality of food products. *Trends in Food Science and Technology* 15:79–85.

Mattila, K., Saris, P., and Työppönen, S. 2003. Survival of *Listeria monocytogenes* on sliced cooked sausage after treatment with pediocin AcH. *International Journal of Food Microbiology* 89:281–286.

Mbandi, E., and Shelef, L.A. 2002. Enhanced antimicrobial effects of combination of lactate and diacetate on *Listeria monocytogenes* and *Salmonella* spp. in beef bologna. *International Journal of Food Microbiology* 76:191–198.

McCormick, K. 2001. In-package pasteurization, combined with nisin-impregnated packaging film to inhibit *L. monocytogenes* and *Salmonella typhimurium* on ready-to-eat meat. MS Thesis, Clemson University, Clemson, SC.

McMullen, L.M., and Stiles, M.E. 1996. Potential for use of bacteriocin-producing lactic acid bacteria in the preservation of meats. *Journal of Food Protection* S9 (Suppl): 64–71.

Ming, X., Weber, G., Ayres, J., and Sandine, W. 1997. Bacteriocins applied to food packaging materials to inhibit *Listeria monocytogenes* on meats. *Journal of Food Science* 62:413–415.

Muriana, P.M., Gande, N., Robertson, W., Jordan, B., and Mitra, S. 2004. Effect of pre-package and post-package pasteurization on post-process elimination of *Listeria monocytogenes* on deli turkey products. *Journal of Food Protection* 67:2472–2479.

Muriana, P.M., Quimby, W., Davidson, C.A., and Grooms, J. 2002. Post-package pasteurization of ready-to-eat deli meats by submersion heating for reduction of *Listeria monocytogenes*. *Journal of Food Protection* 65:963–969.

Murphy, R.Y., and Berrang, M.E. 2002. Effect of steam- and hot-water post-process pasteurization on microbial and physical property measures of fully cooked vacuum-packaged chicken breast strips. *Journal of Food Science* 67:2325–2329.

Murphy, R.Y., Duncan, L.K., Driscoll, K.H., and Marcy, J.A. 2003a. Lethality of *Salmonella* and *Listeria innocua* in fully cooked chicken breast meat products during post-cook in-package pasteurization. *Journal of Food Protection* 66:242–248.

Murphy, R.Y., Duncan, L.K., Driscoll, K.H., Marcy, J.A., and Beard, B.L. 2003b. Thermal inactivation of *Listeria monocytogenes* on ready-to-eat turkey breast meat products during post-cook in-package pasteurization with hot water. *Journal of Food Protection* 66:1618–1622.

Murphy, R.Y., Hanson, R.E., Feze, N., Johnson, N.R., Scott, L.L., and Duncan, L.K. 2005a. Eradicating *Listeria monocytogenes* from fully-cooked franks by using an integrated pasteurization-packaging system. *Journal of Food Protection* 68:507–511.

Murphy, R.Y., Hanson, R.E., Johnson, N.R., Chappa, K., and Berrang, M.E. 2006. Combining organic acid treatment with steam pasteurization to eliminate *Listeria monocytogenes* on fully cooked frankfurters. *Journal of Food Protection* 69(1):47–52.

Murphy, R.Y., Hanson, R.E., Johnson, N.R., Scott, L.L., Feze, N., and Chappa, K. 2005b. Combining antimicrobial and steam treatments in a vacuum packaging system to control *Listeria monocytogenes* on ready-to-eat franks. *Journal of Food Science* 70:138–140.

Nelson, K.A., Busta, F.F., and Sofos, J.N. 1980. Effect of product pH and ingredient forms in chicken frankfurter emulsions on *Clostridium botulinum* growth and toxin production. *Presented at the 40th Annual Meeting of the Institute of Food Technology*, New Orleans, LA, June 8–11.

Nnanna, I.A., Ukuku, D.O., McVann, K.B., and Shelef, L.A. 1994. Antioxidant activity of sodium lactate in meat and model systems. *Lebensmittel-Wissenschaft und-Technologie* 27:78–85.

Ouattara, B., and Mafu, A.A. 2000. Natural antimicrobials in combination with gamma irradiation. In: Roller, S. (ed), *Natural Antimicrobials for the Minimal Processing of Foods*. CRC, Woodhead Publishing, Ltd, Great Abington, Cambridge, UK.

Ouattara, B. Simard, R.E., Piette, G., Begin, A., and Holly, R.A. 2000. Diffusion of acetic and propionic acids from chitosan-based antimicrobial packaging films. *Journal of Food Science* 65(5):768–773.

Palumbo, S.A., and Williams, A.C. 1994. Control of *Listeria monocytogenes* on the surface of frankfurters by acid treatments. *Food Microbiology* 11:293–300.

Papadopoulos, L.S., Miler, R.K., Acuff, G.R., Lucia, L.M., Vanderzant, C., and Cross, H.R. 1991a. Consumer and trained sensory comparisons of cooked beef top rounds treated with sodium lactate. *Journal of Food Science* 56:1141–1146, 1153.

Papadopoulos, L.S., Miller, R.K., Ringer, L.R., and Cross, H.R. 1991b. Sodium lactate effect on sensory characteristics, cooked meat color, and chemical composition. *Journal of Food Science* 56:621–626, 635.

Pietrazak, D., Fonberg-Broczek, M., Mucka, A., and Windyga, B. 2007. Effects of high pressure treatment on the quality of cooked pork ham prepared with different levels of curing ingredients. *High Pressure Research* 27(1):27–31.

Piette, G., and Brodeur, C. 2001. Ohmic cooking for meat products: the heat is on! *Le Monde Alimentaire* 5:22–24.

Poovey, K.A., Burson, D.E., Thippareddi, H., and Mandigo, R.W. 2008. Effects of sodium lactate, sodium citrate, and sodium diacetate on microbiological quality and inhibition of *Listeria monocytogenes* in ready-to-eat hams. *Poster at International Association of Food Protection Annual Conference*, Columbus, OH.

Rhodehamel, E.J., and Beckwith, S.W. 1999. Postprocessing pasteurization of processed meats. *Proceedings of the Reciprocal Meat Conference* 52:113–114.

Roering, A.M., Wierzba, R.K., Ihnot, A.M., and Luchansky, J.B. 1998. Pasteurization of vacuum-sealed packages of summer sausage inoculated with *Listeria monocytogenes*. *Journal of Food Safety* 18:49–56.

Rourke, T.J. 2001. Enhancement of cooked meat quality and safety via packaging. *Proceedings of the Reciprocal Meat Conference* 54:73–76.

Rowan, N.J., MacGregor, S.J., Anderson, J.G., Fouracre, R.A., McIlvaney, L., and Farish, O. 1999. Pulsed light inactivation of food-related microorganisms. *Applied and Environmental Microbiology* 65(3):1312–1315.

Samelis, J., Bedie, G.K., Sofos, J.N., Belk, K.E., Scanga, J.A., and Smith, G.D. 2002. Control of *Listeria monocytogenes* with combined antimicrobials after post-process contamination and extended storage of frankfurters at 4°C in vacuum packages. *Journal of Food Protection* 65:299–307.

Samelis, J., and Sofos, J. 2000. Organic acids. In: Roller, S. (ed), *Natural Antimicrobials for the Minimal Processing of Foods*. CRC, Woodhead Publishing Ltd, Great Abington, Cambridge, UK.

Samelis, J., Sofos, J.N., Kain, M.L., Scanga, J.A., Belk, K.E., and Smith, G.C. 2001. Organic acids and their salts as dipping solutions to control *Listeria monocytogenes* inoculated following processing of sliced pork bologna of sliced pork bologna stored at 48°C in vacuum packages. *Journal of Food Protection* 64:1722–1729.

Scannell, A.G.M., Hill, C., Ross, R.P., Marx, S., Hartmeier, W., and Arendt, K. 2000. Development of bioactive food packaging materials using immobilized bacteriocins Lacticin 3147 and Nisaplin. *International Journal of Food Microbiology* 60:241–249.

Seman, D.L. 2001. Safety and quality concerns – ingredients. *Proceedings of the 54th Reciprocal Meat Conference* 54:68–72.

Seman, D.L., Borger, A.C., Meyer, J.D., Hall, P.A., and Milkowski, A.L. 2002. Modeling the growth of *Listeria monocytogenes* in cured, ready-to-eat processed meat products by manipulation of sodium chloride, sodium diacetate, potassium lactate, and product moisture control. *Journal of Food Protection* 65:651–658.

Seydim, A.C., and Sarikus, G. 2006. Antimicrobial activity of whey protein based edible films incorporated with oregano, rosemary and garlic essential oils. *Food Research International* 39(5):639–644.

Shelef, L.A. 1994. Antimicrobial effects of lactates: A review. *Journal of Food Protection* 57(5):445–450.

Shelef, L.A., and Yang, Q. 1991. Growth suppression of *Listeria monocytogenes* by lactates in broth, chicken, and beef. *Journal of Food Protection* 54(4):283–287.

Singh, A., Singh, R.K., Bhunia, A.K., and Singh, N. 2003. Efficacy of plant essential oils as antimicrobial agents against *Listeria monocytogenes* in hotdogs. *Lebensmittel-Wissenschaft und-Technologie* 36(8):787–794.

Sofos, J.N., and Busta, F.F. 1980. Antimicrobial activity of sorbate. *Journal of Food Protection* 44(8):614–622.

Stekelenberg, F.K., and Kant-Muermans, M.L.T. 2001. Effects of sodium lactate and other additives in a cooked ham product on sensory quality and development of a strain of *Lactobacillus curvatus* and *Listeria monocytogenes*. *International Journal of Food Microbiology* 66:197–203.

Suzuki, A., Kim, K., Tanji, H., and Nishiumi, T. 2006. Application of high hydrostatic pressure to meat and meat processing. In: Nollet, L.M.L., and Toldra, F. (eds), *Advanced Technologies for Meat Processing*. Taylor & Francis Group, New York.

Tantillo, M.G., DiPinto, A., and Novello, L. 2002. Bacteriocin-producing *Lactobacillus sake* as starter culture in dry sausages. *Microbiology* 25:45–49.

Tassou, C.C., Drosinos, E.H., and Nyghas, G.-J.E. 1995. Effects of essential oil from mint (*Mentha piperita*) on *Salmonella enteritidis* and *Listeria monocytogenes* in model food systems at 4 and 10°C. *Journal of Applied Bacteriology* 78:593–600.

Thippareddi, H., Juneja, V.K., Phebus, R.K., Marsden, C.J.L., and Kastner, C.L. 2003. Control of *Clostridium perfringens* germination and outgrowth by buffered sodium citrate during chilling of roast beef and injected pork. *Journal of Food Protection* 66(3):376–381.

Ting, W.T.E., and Deibel, K.E. 1991. Sensitivity of *Listeria monocytogenes* to spices at two temperatures. *Journal of Food Safety* 12:129–137.

Tompkin, R.B. 1978. The role and mechanism of the inhibition of *Clostridium botulinum* by nitrite – Is a replacement available? *Proceedings of the 31st Red Meat Conference*, Storrs, CT, June 18–22, pp. 135–147.

Unda, J.R., Molins, R.A., and Walker, H.W. 1991. *Clostridium sporogenes* and *Listeria monocytogenes*: Survival and inhibition in microwave-ready beef roasts containing selected antimicrobials. *Journal of Food Science* 56(1):198–205, 219.

USDA FSIS. 2003. Control of *Listeria monocytogenes* in ready-to-eat meat and poultry products; final rule. *Federal Register* 68:34207–34254.

USDA FSIS. 2006. *Compliance Guidelines to Control Listeria monocytogenes in Post-Lethality Exposed Ready-to-Eat Meat and Poultry Products*. http://www.fsis.usda.gov/oppde/rdad/FRPubs/97-013F/LM_Rule_Compliance_Guidelines_May_2006.pdf.

USDA FSIS. 2007. Directive 7120.1, *Safe and Suitable Ingredients Used in the Production of Meat and Poultry Products*.

Valderrama, W.B., Mills, E.W., and Cutter, C.N. 2008. Efficacy of chlorine dioxide against *Listeria monocytogenes* in brine solutions. In: *Abstracts of the Annual Meeting of the International Association for Food Protection (P2-39)*, pp. 89–90.

Venkitanarayanan, K.S., Ezeike, G.O., Hung, Y.C., and Doyle, M.P. 1999. Efficacy of electrolyzing water for inactivating *E. coli* O157:H7, *Salmonella enteriditis*, and *Listeria monocytogenes*. *Applied and Environmental Microbiology* 65(9):4276–4279.

Vicente, A.A., de Castro, I., and Teixeira, J.A. 2006. Ohmic heating for food processing. In: Sun, D.-W. (ed), *Thermal Food Processing: New Technologies and Quality Issues*. CRC–Taylor & Francis Group, Abingdon, Oxfordshire, UK.

Wagner, M.K., and Busta, F.F. 1983. Effect of sodium acid pyrophosphate in combination with sodium nitrite or sodium nitrite/potassium sorbate on CBot growth and toxin production in beef/pork frankfurter emulsion. *Journal of Food Science* 48:990–991, 993.

Wang, W.C., Chen, J.I., and Hua, H.H. 2002. Study of liquid-contact thawing by ohmic heating. *Paper presented in Proceedings of the IFT Annual Meeting and Food Expo*, Anaheim, CA.

Yen, L.C., Sofos, J.N., and Schmidt, G.R. 1991. Effect of meat curing ingredients on thermal destruction of *Listeria monocytogenes* in ground pork. *Journal of Food Protection* 54(6):408–412.

Yousef, A.E., and Marth, E.H. 1988. Inactivation of *Listeria monocytogenes* by ultraviolet energy. *Journal of Food Science* 53(2):571–573.

CHAPTER 6

Introduction to Lethality Equations

Bradley P. Marks, Michigan State University

Introduction

As noted elsewhere in this book, *Salmonella* is the target pathogen for regulations governing thermal processing of ready-to-eat (RTE) meat and poultry products. Specific requirements are prescribed by the USDA-FSIS lethality performance standards for RTE products (FSIS, 1999, 2001a, 2001b).

The amended regulations state that any process producing RTE, whole-muscle products must achieve 6.5-$\log_{10}$ or 7.0-$\log_{10}$ reduction in *Salmonella* for beef or poultry, respectively. A proposed rule (FSIS, 2001a) would extend these standards to all meat and poultry products. The rule changes do allow for processing according to specified "safe harbor" conditions; however, these are available for only a limited number of products (FSIS, 2001b), so that a wide variety of products need independent analysis of process lethality. Therefore, predictive microbial models are extremely valuable tools in validating that a given process is compliant with these requirements.

As noted in an FSIS notice (FSIS, 2002), predictive models alone are insufficient for ensuring process validation. The model inputs, and application of the model outputs, require expert (microbiological) interpretation, and a separate chapter (see Chapter 12) highlights various factors that can affect the validity of model predictions. Nevertheless, lethality models are increasingly important tools for process validation, given the impracticality/impossibility of conducting experimental challenge studies with *Salmonella* in actual processing facilities.

Therefore, the goal of this chapter is to introduce the reader to a very basic methodology for calculating thermal inactivation of *Salmonella* (i.e., process lethality) for RTE meat and poultry products. It is *not* the goal of this chapter to cover the broad body of work associated with microbial modeling, or to cover various alternative model forms. Additionally, this author is not asserting that the methods presented in this chapter encompass the *best* model form (in terms of fundamental principles or phenomenological basis); rather, the method being presented is that which is widely used within the industry and well accepted by the relevant regulatory personnel. It is therefore the one most likely to be encountered by the reader (at this point in time); consequently, this chapter is designed to present the methods in a very simple and applied format.

The reader who is interested in more detailed information on microbial models is directed to several comprehensive books on the subject by McKeller and Lu (2004), McMeekin et al. (1993), and Peleg (2006).

Therefore, the specific objectives of this chapter are to define key terminology associated with microbial models, outline the basic principles of log-linear models, and illustrate a step-by-step process for computing process lethality for a given cooking process.

Terminology

To ensure the maximum utility of this chapter, it is critical that the following terms are well understood by the reader: logarithm, D value, z value, and F value. The model terms (D, z, F) will be defined here only in lay language; the specific formulas necessary to calculate them will be presented in the subsequent section.

Logarithm

A logarithm (log) is the exponent to which a base number must be raised to produce a given number. Specifically, $\log_x(X^y) = y$. By example, $\log_{10}100 = 2$, meaning that the base-10 log of 100 (i.e., 10^2) is 2. Similarly, $\log_{10}(1{,}000{,}000) = 6$, $\log_{10}(1{,}000) = 3$, $\log_{10}(10) = 1$, and $\log_{10}(1) = 0$. Therefore, on a base-10 log scale, a "1-log reduction" means a 90% reduction from the original value. Note that bases other than 10 can be used; another common base is known as the natural logarithm, which is typically written as $\ln(x)$, and is equivalent to $\log_e(x)$, where $e \approx 2.718$. This form is occasionally used in microbial modeling literature; therefore, the reader needs to be careful to know which form is used before utilizing published model parameters. In most cases, and in the relevant regulatory language, base-10 log is standard.

D Value

The decimal reduction (D) value is the time required, at a constant, lethal condition (e.g., temperature), to achieve a 1-log (90%) reduction in the initial microbial population.

Z Value

The z value is the change in temperature required to cause a 1-log (10-fold) change in the D value. For example, if $D_{150^\circ\text{F}} = 4.0$ minutes and $z = 10^\circ\text{F}$, then $D_{160^\circ\text{F}} = 0.4$ minutes.

F Value

The F value is the process time at a reference temperature that would achieve equivalent lethality as a given actual process. It is generally used to reduce a nonisothermal process to a single value describing how much time would be required at a fixed reference temperature to achieve the equivalent lethality. The F value is a hold-over from the initial application of microbial models to canning (sterilization) calculations. It is not necessary for calculating thermal lethality for meat cooking processes. However, a definition and description is included in this chapter, because certain software packages report this parameter; therefore, it is important for the reader to know how to interpret and use this value.

Modeling Basics

The general method for computing process lethality is based on the following principles. It is assumed that if a bacterial population is subjected to a constant lethal temperature, the population decreases logarithmically

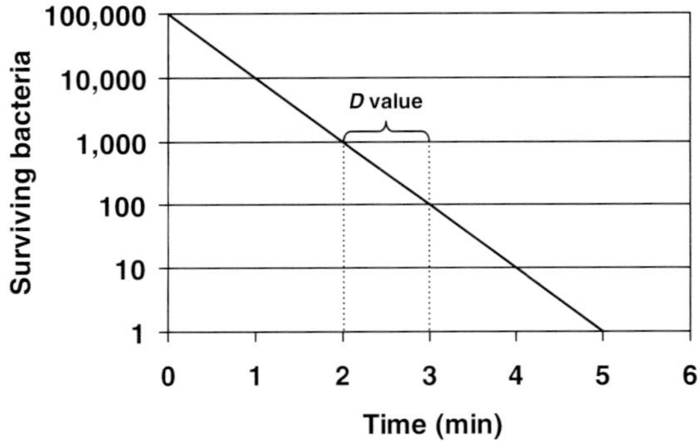

Figure 6.1. An illustration of a log-linear, isothermal inactivation curve, showing the D value as the time required for a 1-log (90%) reduction in bacterial population.

with time; this is commonly referred to as a log-linear model, because a plot of log(population) versus time is a straight line (Fig. 6.1).

Mathematically, this concept is expressed in log reductions as:

$$\log_{10}(N_0) - \log_{10}(N_t) = \frac{t}{D_T} \tag{6.1}$$

which can be rewritten as:

$$\log_{10}\left(\frac{N_t}{N_0}\right) = -\frac{t}{D_T} \tag{6.2}$$

where N_0 is the initial bacterial population, N_t is the surviving population at time t, and D_T is the decimal reduction (D) value at a constant lethal temperature (T). Calculating D_T at a given temperature requires a known D_{ref} at a reference temperature (T_{ref}) and the z value, such that:

$$D_T = D_{\text{ref}} \times 10^{(T_{\text{ref}}-T)/z} \tag{6.3}$$

A separate chapter (see Chapter 12) discusses the fact that the thermal resistance of *Salmonella* (and therefore D) is affected by many factors other than just temperature, such as fat content, moisture content, and product structure. However, the equation above accounts only for the temperature effect. Therefore, the user needs to be sure that the D_{ref} and z values being used are appropriate for the product/process being evaluated.

Lastly, the F value, previously defined, can be calculated as:

$$F_{\text{ref}} = D_{\text{ref}} \times \text{log reductions} \tag{6.4}$$

Table 6.1. Sample lethality calculations for a simplified example process, where $D_{ref} = 12$ seconds , $T_{ref} = 145°F$, and $z = 10°F$.

Time (s)	*Temperature (°F)*	$D_T = D_{ref} \times 10^{(T^{ref} - T)/z}$ *(s)*	$-\Delta t / D_T$
0–30	145	120	0.25
30–50	155	12	1.67
50–60	165	1.2	8.33
		Cumulative lethality = 10.25 log reductions	

or, if software is computing F values, then log reductions can be calculated from:

$$\text{log reductions} = \frac{F_{ref}}{D_{ref}} \tag{6.5}$$

noting that an F value is always at a specified reference temperature.

Obviously, no meat cooking process results in a constant product temperature. Therefore, in real applications, the above concept needs to be applied to a continuously varying temperature profile, where D_T changes with time. This is done by integrating Eq. (6.2) over time to calculate cumulative log reductions. This will be illustrated here via several examples, starting with the simplest case.

For the first case, consider a process (fictional) where a meat product is instantly heated to a core temperature of 155°F and held for 30 seconds. If $D_{155°F}$ for *Salmonella* in this product is known to be 0.2 minutes (12 seconds), then the lethality for this process is (30 seconds)/(12 seconds) = 2.5 log reductions.

For the second case, consider a slightly more complicated temperature history, where the product core temperature is at 145°F for 30 seconds, then 155°F for 20 seconds, and then 165°F for 10 seconds. Then, the cumulative lethality is the sum of lethalities for each time step. If z for *Salmonella* in this product is known to be 10°F, Table 6.1 illustrates the lethality calculation for this process.

Implementation of these calculations in a spreadsheet format will be illustrated in the following section.

Lethality Calculations

Computing *Salmonella* lethality for an actual thermal process entails several steps. First, experimental data must be collected for the core temperature of the product during the thermal process. Then, appropriate model parameters must be acquired. Lastly, those model parameters must be applied to the temperature data to calculate a cumulative lethality. These steps are described in more detail below.

Experimental Data

The quality of the computed lethality can be no better than the quality of the data to which the model is applied. Therefore, it is critically important to collect high-quality temperature data from the cold spot in the product during cooking. An appropriate data logger, with calibrated temperature probes, should be used to collect core temperature data as the product is subjected to the desired process. The temperature probe should be fixed sufficiently well to ensure that the measuring tip remains in the expected cold spot even as the product is processed (particularly if it is traveling through a continuous oven system).

The minimum number of time–temperature data points collected for a process should be 30 or more; however, given that data capacity is rarely a concern with current data logger systems, additional data provide a more complete description of the temperature history of the product. Additionally, it is recommended that temperature profiles be collected from replicate samples (≥ 3) that represent expected variability in the process (e.g., from different positions across a belt), so that the potential variability in lethality can be quantified.

Model Parameters

Before the inactivation model can be applied to the collected temperature data, the user must acquire D_{ref}, T_{ref}, and z for the product and process being validated. Chapter 12 includes numerous references that include these parameters (or other model parameters) for *Salmonella* in various meat products. Additionally, ComBase is an international microbiological database (www.combase.cc) that serves as a repository of food microbiology data, including data for *Salmonella* lethality in meat products. The user is encouraged to find inactivation parameters that best match the product and process being validated. Additionally, when utilizing inactivation parameters from ComBase or any other source, the user should acquire and review the original document from which the parameters came, to ensure that the model parameters are validated and suitable; the original reference should be retained in the records with the lethality calculations. Unfortunately, it is generally very difficult to find inactivation parameters that exactly match one's product and process; therefore, processors should be very cautious in utilizing results based on model parameters that are extrapolated from a different product type or composition.

Spreadsheet Implementation

The modeling concepts outlined above can be implemented easily via a spreadsheet application. Figure 6.2 illustrates a simple example of such an implementation. The only difference from the prior examples is that the simple examples (e.g., Table 6.1) had a constant temperature over each time step. In the real case, temperature varies during each time step; therefore, after D_T is calculated at each measured point (based on Eq. (6.3)), a $(1/D_T)_{avg}$ value must be calculated for each time step, based on the computed $1/D_T$ values for the beginning and end of each time step. Then, the $(1/D_T)_{avg}$ value for each time step is multiplied by the size of the time step, and a running total is computed in the final column. The

Computing thermal process lethality using the general method

T_{ref} =	145	°F
D_{ref} =	42	s
z =	10.3	°F

t (s)	T (°F)	D_T (s)	$1/D_T$ (1/s)	$(1/D_T)_{avg}$ (1/s)	Cumulative Lethality $-\Sigma[\Delta t \times (1/D_T)_{avg}]$
0	25.3	1.8E+13	5.7E–14	–	0
120	28.7	8.2E+12	1.2E–13	8.9E–14	–1.07E–11
135	39.9	6.7E+11	1.5E–12	8.1E–13	–2.28E–11
150	58.9	9.6E+09	1.0E–10	5.3E–11	–8.15E–10
165	76.7	1.8E+08	5.6E–09	2.8E–09	–4.34E–08
180	92.8	4.9E+06	2.0E–07	1.0E–07	–1.61E–06
195	106.5	2.3E+05	4.4E–06	2.3E–06	–3.58E–05
210	120.4	1.0E+04	9.7E–05	5.1E–05	–7.99E–04
225	130.2	1.1E+03	8.7E–04	4.8E–04	–8.06E–03
240	141.4	9.4E+01	1.1E–02	5.8E–03	–0.09
255	150.9	1.1E+01	8.9E–02	5.0E–02	–0.84
270	160.1	1.4E+00	7.0E–01	3.9E–01	–6.73

Figure 6.2. An illustration of a spreadsheet for computing cumulative process lethality, given model parameters (T_{ref}, D_{ref}, and z) and process time–temperature data.

example process in the figure resulted in a cumulative process lethality of 6.73 log reductions.

A similar spreadsheet is freely available from the American Meat Institute Foundation (http://www.amif.org). The AMI Process Lethality Spreadsheet (AMI-PLS) includes graphs of the product temperature and process lethality. However, AMI-PLS computes an F value for the process, rather than log reductions; therefore, the user must calculate the log reductions as previously described (Eq. (6.5)).

Summary

Microbial inactivation models are useful and important tools for evaluating process safety and regulatory compliance. However, all lethality calculations have some uncertainty associated with them, even though it is rarely well known or reported. For example, the previous example illustrated a process resulting in 6.73 log reductions. In reality, the underlying uncertainty in the original model parameters may have meant that the actual process lethality was 6.73 ± 0.8 log reductions (i.e., with a 95% confidence interval of 5.93–7.53 log reductions). If this process was for a beef product, with a lethality performance standard of 6.5 log reductions, then the user cannot be certain that 6.73 is actually sufficient.

Therefore, it is important for processors to (1) know as much as possible about the data and parameters that are being used in the model, matching them as closely as feasible to the product and process being validated, (2) document the information sources and data collection methods, and (3) apply the results cautiously. Predictive models are powerful tools, when used judiciously and in an informed manner, and are an important part of process validation and regulatory compliance.

References

FSIS. 1999. Performance standards for production of certain meat and poultry products. United States Department of Agriculture. Food Safety Inspection Service. 9 CFR Parts 301, 317, 318, 320, and 381. *Federal Register* 64(3):732–749.

FSIS. 2001a. Performance standards for the production of processed meat and poultry products; proposed rule. United States Department of Agriculture. Food Safety Inspection Service. 9 CFR Parts 301, 303, et al. *Federal Register* 66(39):12590–12636.

FSIS. 2001b. *Draft Compliance Guidelines for Ready-to-Eat Meat and Poultry Products*. United States Department of Agriculture. Food Safety Inspection Service. http://www.fsis.usda.gov/OPPDE/rdad/FRPubs/RTEGuide.pdf.

FSIS. 2002. *Use of Microbial Pathogen Computer Modeling in HACCP Plans*. FSIS Notice 55–02. United States Department of Agriculture. Food Safety Inspection Service, December 2, 2002.

McKellar, R., and Lu, X. 2004. *Modeling Microbial Responses in Foods*. CRC Press, Boca Raton, FL.

McMeekin, T.A., Olley, J.N., Ross, T., and Ratkowsky, D.A. 1993. *Predictive Microbiology Theory and Applications*. John Wiley & Sons, New York.

Peleg, M. 2006. *Advanced Quantitative Microbiology for Foods and Biosystems: Models for Predicting Growth and Inactivation*. CRC Press, Boca Raton, FL.

CHAPTER 7

Predictive Microbiology Information Portal with Particular Reference to the USDA—Pathogen Modeling Program*

Vijay Juneja,[1] Cheng-An Hwang,[1] and Mark Tamplin[2]
[1]United States Department of Agriculture, Agricultural Research Service, Eastern Regional Research Center, Microbial Food Safety Research Unit.
[2]School of Agricultural Science, University of Tasmania, Australia.

Predictive Microbiology Information Portal

To ensure the safety of the nation's food supply, the Food and Drug Administration (FDA) and the United States of Department of Agriculture (USDA) establish food safety regulations that govern the manufacturing and distribution of domestic and imported food products. Food products must meet these regulations so they can be marketed in the US. However, these regulations are often not understood even by large food companies, albeit that they have financial resources to employ food safety experts to interpret and help them comply with the regulations. Small and very

* Mention of trade names or commercial products in this chapter is solely for the purpose of providing specific information and does not imply recommendation or endorsement by the U.S. Department of Agriculture.

small food producers are often unable to understand and comply with food safety regulations, which result in limited production volume or close of business and lead to financial hardship. Therefore, the Predictive Microbiology Information Portal (PMIP) was developed to assist food processing companies, particularly small and very small ones, in the use and interpretation of food safety regulations, microbiological predictive models, and microbial growth data. The PMIP is a comprehensive web site that brings together food safety policies, guidelines and regulations, pathogen predictive models, microbiological research data, and useful food safety resources. This internet-based food safety information portal can assist food processors in achieving a high level of safety, quality, and wholesomeness of their products. The portal was open to the public in 2007 and can be accessed at http://portal.arserrc.gov/. Users, once access to the portal, can quickly find food safety regulations through a user-friendly, easy-to-use interface to retrieve information of their interests. They can access food safety regulations, information resources, microbiological models, and microbial growth database to estimate the safety of their products. The portal brings food regulations and tens of thousands of microbiological data to the fingertips of the users. The users not only utilize the portal to comply with food safety regulations but also to reduce costs due to noncompliance product recalls. In summary, the portal is meant for easy and rapid access to relevant food safety information and is a tool for making timely and accurate food safety decisions.

For the food industry, the PMIP is a useful tool for locating and retrieving predictive models, regulations, and research data for use in validating critical control points and corrective actions in HACCP systems. It reduces food safety challenge studies needed in determining quality and safety of new products.

The portal provides a searchable function that users can use to obtain specific information that is of interest to them. The tutorial section provides brief instructions and examples on how to navigate the portal and retrieve information. PMIP contains three major sections that provide an access to predictive models for foodborne pathogens, regulatory policies and guidelines, and microbial data related to pathogenic and spoilage microorganisms in a wide variety of food products. The predictive models are mainly from the Pathogen Modeling Program (PMP) of the Eastern Regional Research Center (ERRC), Agricultural Research Service (ARS), US Department of Agriculture (USDA). The main sources of rules/regulations are links to the Food Safety Inspection Service, USDA and the Food and Drug Administration (FDA) web sites. The microbial data are from ComBase, the world's largest online relational database of predictive microbiology information, which contains about 40,000 microbial data records that describe the growth, survival, and inactivation of bacteria under diverse environments relevant to food products and food processing operations. ComBase is jointly developed and maintained by the Food Research Institute of UK, ARS-USDA, and the Center of Excellence for Food Safety, Australia. Users can enter the food formulation parameters of their interests and retrieve microbial growth data that match their search criteria. The use of ComBase eliminates conducting unnecessary microbiological studies, reduces redundancy of laboratory experiments, increases researcher's efficiency, and, as a result, enhances the safety of foods. The internet version of ComBase browser is hosted on a server at the ERRC-ARS-USDA. Researchers at the ERRC have the charge to maintain and modify the browser, to improve its functionality and efficiency in response to users input, and to update with new version of the ComBase database. This chapter focuses on the use of the USDA-PMP.

Pathogen Modeling Program

The food industry requires information on how the growth, survival, and inactivation of pathogenic bacteria are influenced by different intrinsic and extrinsic environmental conditions, such as salt, pH, temperature, atmosphere, and preservatives, in foods to ensure the safety of their products. PMP, developed by research scientists at the ERRC-ARS-USDA, consists of various models that describe the effect of these environmental conditions on the behavior of common foodborne pathogenic and spoilage microorganisms. The scientists at the ERRC organized and conducted extensive microbiological experimentation with foodborne pathogens. Microbial growth, survival, or inactivation data were developed into primary mathematical models to describe microbial changes with time. Secondary mathematical models were then developed to describe the effects of environmental factors on the microbial growth, survival, or inactivation. These models can be used to predict the growth of most common pathogens as affected by environmental conditions. Realizing that these models in mathematical equations would not be fully understood by users, the scientists transformed these predictive equations into an easy-to-use computer software, namely, the PMP. It is free for download from the ARS web site or can be accessed online. This program does not require any modeling and mathematical knowledge or calculations by the users. PMP allows the users to input environmental criteria and then to retrieve predictions about how pathogenic bacteria react to specific food environments. To further assist food processors in meeting regulatory requirements, references are provided for each model via direct internet access to portable document format (PDF) files containing the research articles with the developed models. The models enable food processors to assess the microbial risks of a particular food and to estimate consequences of out-of-control process events, for example, cooling deviation and refrigeration failure.

Modeling of Microbial Pathogens in Foods

The theory of predictive microbiology is based on the fact that the microbial growth, survival, and inactivation are affected by the environmental factors. It is also based on the assumption that the responses of microorganisms to these factors are reproducible and can be characterized and quantified. Therefore, the microbial response to environmental factors can be described mathematically. Microbiological modeling is an attempt to define the response of a microorganism to its environment in terms of mathematical equations.

Knowing the fate of foodborne pathogens in foods is important in determining the microbiological safety levels of food products. When the potential growth of a foodborne pathogen is likely in a food product, the food product then needs to be either reformulated or processed to eliminate the presence or growth potential of pathogens to ensure food safety. The growth, survival, or death of microorganisms in food is affected by many factors, both intrinsic and extrinsic. These include intrinsic factors such as pH, NaCl (salt), sugars, phosphates, nitrites, water activity, and nutrient level, and extrinsic factors such as temperature, atmosphere (i.e., aerobic, anaerobic, and modified atmosphere), and relative humidity. By manipulating one or more of these factors, it is possible to alter the growth behavior of microorganisms in foods, for examples, extending lag and generation time of microorganisms, increasing sensitivity of microbial cells or spores to intervention technology, or preventing spore outgrowth and toxin production. By quantifying the effects and interactions of such multiple food factors, predictive mathematical models can be developed to predict growth behavior of microorganisms in foods. Steps for development of predictive models include experimental design, data collection, model development, model validation, and development of an effective interface between the models and the end user. An example of the interface is the PMP developed by the USDA.

PMP is a computer-based software program that constitutes a group of models that can be used to estimate the behavior of common foodborne pathogens in various environmental conditions. Through the user-friendly interface, users can select models of their interest and input environmental conditions to retrieve predictions on the growth, inactivation/lethality (thermal and nonthermal), and decline/survival of microorganisms. The models in PMP cover almost all the common foodborne pathogens in foods and environmental factors found in common types of food products. The PMP is a useful tool for obtaining microbial growth potential to be used in designing HACCP plans, in identifying critical control points and critical limits, and in evaluating the consequences of process deviations as well as in determining safe corrective actions to be taken.

Before using the PMP, the user should become familiar with some basic knowledge regarding the microbial growth, principles and the development of predictive models, and the limitation of their uses. The latter is particularly important when applying information obtained from the models to real-world use. The users should recognize that the predictions obtained from PMP models cannot be solely relied on as a sole means of ensuring the microbiological safety of food products. The models cannot replace microbial validation or experimental challenge studies. Models in PMP were developed from studies that closely simulated the intrinsic and extrinsic environmental conditions that are relevant to the food products of interest. However, not all the conditions can be completely incorporated into the studies to represent the real-world food systems. The predictions obtained from PMP, therefore, cannot be guaranteed to indicate the microbial behavior in food systems. The following describes some basic knowledge regarding microbiological growth and inactivation that the user should be familiar with.

Phases of Bacterial Growth

The level of microorganisms in food is controlled by various factors, including the initial contamination level, the level of nutrients, temperature, pH, water activity, additives, and the presence of other microorganisms. In food, microorganisms can increase in numbers (grow), decrease in numbers (inactivate or die), or remain at the same level (survive). Predictive models can be developed for each of these types of bacterial behavior. Microbial growth can be segmented into three different phases: lag phase, growth phase, and stationary phase. A graphical representation of the bacterial growth cycle is depicted in Fig. 7.1.

Lag Phase (Lag Phase Duration)

Lag phase can be defined as the time required for the cell population to adjust to the food environment or a new environment prior to replication

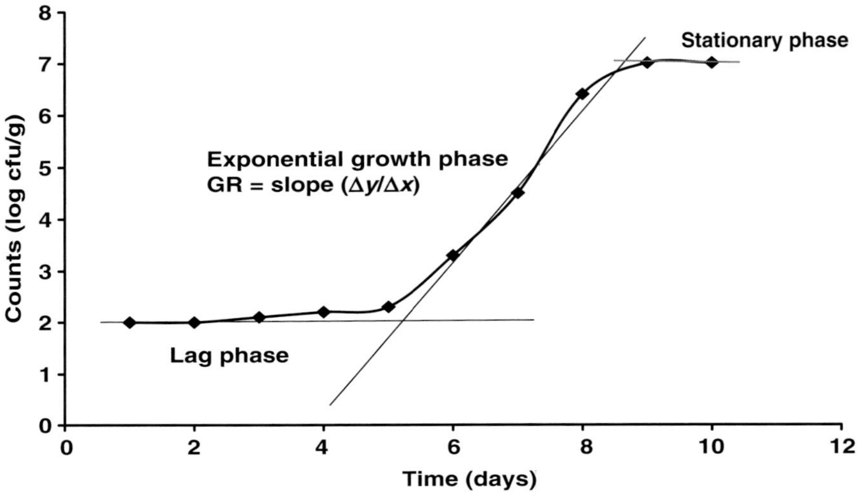

Figure 7.1. Phases of microbial growth.

(growth). Lag phase is the most unpredictable part of a growth curve compared to growth and stationary phases. This is because lag phase is not only dictated by the innate properties of the cell but also by the previous physiological state (history) of the microorganism. For example, the lag phase duration (LPD) of bacteria grown at 37°C (98°F) in culture media and then transferred to raw ground beef at 10°C (50°F) will be different than the LPD of bacteria grown at 21.1°C (70°F) and then transferred to ground beef at 10°C (50°F). This is because the previous environment of the bacteria will result in different cellular constituents that need to be made before the growth commences in a new environment. LPD was reported to be higher in brain heart infusion broth as compared to the values in sterile raw ground beef (Tamplin et al., 2005). Thus, the predictions from the models developed from the data collected in broth cannot be applied to another substrate.

Growth Phase

The growth phase represents the replication (multiplication) of microorganisms. The cell numbers are increasing at such a rate that it is best to use logarithm values to represent them graphically. Growth is sometimes described in terms of growth rate or generation (doubling) time. Growth rate is expressed as the change in microbial population per unit time. Mathematically, growth rate, termed as "specific growth rate" (h^{-1}), is generally expressed in natural logarithm (ln) form. Specific growth rate is obtained by multiplying $\log_{10}$ form of growth rate by the ln (10) or ~2.303. The generation time is the time (usually stated in hours or days) that it takes for one cell to divide and become two cells. The generation time is referred to as the doubling time for the entire population. In a population of a bacterial species, not all cells divide at the same time or at the same rate. Generally, bacteria have longer generation times in food systems than in bacteriological broth medium. To convert generation time to growth rate, simply divide 0.301 (the $\log_{10}$ value of 2) by the generation time. On the other hand, growth rate is the change in bacterial numbers over some period of time, typically expressed as $\log_{10}$ per hour or day. To convert growth rate to generation (doubling) time, divide 0.301 by the growth rate.

Stationary Phase and Maximum Population Density

Stationary phase is the phase that microbial growth ceases and the number of cells reaches the maximum population density (MPD). Microbial growth ceases due to the presence of other bacteria, exhaustion of nutrients, or the production of inhibitory factors such as toxic metabolites, low pH, and depletion of essential growth factors (Jay, 2000). In most foods, a typical MPD is 9–10 $\log_{10}$ (1 billion to 10 billion) cells per gram or milliliter of food.

Death Phase

Bacteria die in a food after reaching the stationary phase due to the depletion of nutrients and the presence of toxic metabolites produced during growth phases.

Phases of Bacterial Inactivation

Bacteria are inactivated or killed when environmental conditions are adverse to bacterial survival. These conditions can cause acute (fast) inactivation as with high temperature, or mild inactivation (slow), as observed with low levels of organic acids. The shape of the inactivation curve may vary, depending on the organism and the environmental conditions. Conditions may cause an immediate death or linear reduction of bacterial cells, or a period of no death followed by a graduate death of bacterial cells.

Linear Inactivation Phase

For inactivation scenarios, the $\log_{10}$ value of the cell number versus time is plotted. In the linear phase of inactivation, the rate (slope of the line) of inactivation depends on the severity of the effectors (such as heat). Thermal inactivation is commonly referred to in terms of the decimal reduction time, or *D* value. It is the time for the microbial population to decrease by 90% (10-fold or 1.0 $\log_{10}$) at a specific heating temperature. The *D* value is the absolute value of the inverse of the rate (slope) of cell reduction when cell reductions were plotted against heating. Another common term related to thermal inactivation is the *z* value. It describes the change in heating temperature that causes a 90% (or 10-fold) reduction in the *D* value. The *z* value is commonly used to calculate process lethality (*F* value). Process lethality can be expressed as the *F* value, which is an integrated calculation of time-dependent thermal effects on inactivation of cell numbers, and serves to measure the accumulated lethality effects during a complete thermal process, which includes “come-up” and “come-down” times.

Shoulders and Tails

The kinetics of both thermal and nonthermal inactivation may display a lag-like period, sometimes referred to as a “shoulder,” that proceeds the linear inactivation phase. For thermal inactivation scenarios, this is more commonly observed at lower temperatures and when using higher cell concentrations. It has been theorized that this represents a subpopulation of cells that are more thermotolerant, which is more likely to be observed at high-inoculum levels. Shoulders may result from inaccurate measurements of the internal temperature of the matrix during temperature “come-up” time, the use of mixed cultures, cell clumping, and cell multiple-hit mechanisms. The Weibull distribution is commonly applied to model, such as nonlinear inactivation curves (Albert and Mafartb, 2005; Chen, 2007; Virto et al., 2006). In some instances, the linear phase of inactivation does not intercept the *x*-axis, but instead transitions to a curve referred to as a “tail.” Such “tails” are more commonly observed with higher inoculum levels. It has been theorized that “tails” represent a subpopulation of bacteria that are more thermally resistant.

Classes of Models

Primary Models

After the experimental protocol is established and experiments are conducted, cell number versus time data are collected for each of the test conditions. The data are fitted and analyzed by curve-fitting programs to develop a best-fit line to the data to obtain parameters of microbial growth characteristics. For growth data, these may include LPD, growth rate, and MPD for the microorganism. For inactivation data, these may include an initial “shoulder,” a linear reduction in cell count (inactivation rate), and possibly a “tail.” For probability-of-growth or growth/no-growth boundaries, growth or no-growth data are obtained.

Secondary Models

Secondary models are derived from parameters (e.g., lag time, growth/inactivation rate, and MPD) obtained from the primary model. Secondary models describe the primary model parameters as a function of the environmental factors that are examined in the experimental studies. For example, a secondary model describes the growth rates of a pathogen in a food product as a function of storage temperature, or growth rates as a function of salt concentration, water activity, storage temperature, etc. In thermal inactivation of microorganisms, the *z* value is a type of secondary model that describes the change in *D* value as a function of temperature. Secondary models can be simple linear regressions or complex polynomial models.

Various secondary models have been used to describe growth and inactivation of bacteria. Lag time, growth rate, and MPD of microbial growth have been described using square-root, gamma, and cardinal approaches (Le Marc et al., 2005). There are also probability models to describe the likelihood of a microbial event to occur in foods, such as modeling growth/no-growth interfaces, and the production of microbial toxins.

Tertiary Models

The next step of model development involves presenting the predictions obtained from the secondary models in a primary model fashion (cell number vs. time). Tertiary models are the interface between the models and the user and predict the growth/inactivation of a specific or group of microorganisms within the range of experimental conditions. These interfaces are commonly done with computer software such as spreadsheets (e.g., Microsoft Excel), and programming language (e.g., Visual Basic and C++). An example of such interface is the USDA-ARS's PMP (http://ars.usda.gov/Services/docs.htm?docid = 6786). PMP is developed by a research group at the USDA-ARS-ERRC in Wyndmoor, PA. The PMP contains models that can be used to predict the growth and inactivation of foodborne bacteria, primarily pathogens, under various environmental conditions that are relevant to food products. These predictions are specific to the bacterial strains and environmental conditions that were used to generate the models. The PMP has been distributed in various forms, ranging from spreadsheets to stand-alone software, and most recently available online.

USDA-ARS-PMP

Types of PMP Models

Growth Models

The majority of PMP models are growth models. Model variables include atmosphere (aerobic and anaerobic), temperature, pH, water activity, and additives such as nitrite and phosphates. Aerobic growth models are for *Aeromonas hydrophila*, *Bacillus cereus*, *Clostridium perfringens*, *Escherichia coli* O157:H7, *Listeria monocytogenes*, *Salmonella* spp., *Shigella flexneri*, *Staphylococcus aureus*, and *Yersinia enterocolitica*. Anaerobic growth models are for *A. hydrophila*, *B. cereus*, *C. perfringens*, *E. coli* O157:H7, *L. monocytogenes*, *S. flexneri*, and *S. aureus*.

Survival (Nonthermal Inactivation) Models

The survival models predict the inactivation of bacterial pathogens as a function of temperature, NaCl, pH, nitrite, and lactic acid. The publication for each model should be read to determine the acid that was used in the broth models to adjust broth pH. In general, survival models were developed using an organic acid (lactic acid) as the acidulant. Nonthermal inactivation/survival models are for *E. coli* O157:H7, *L. monocytogenes*, *Salmonella* spp., and *S. aureus*.

Thermal Inactivation Models

There are three thermal inactivation models. They are *C. botulinum*, *E. coli* O157:H7, and *L. monocytogenes*, with variables for temperature, pH, NaCl, and sodium pyrophosphate.

In their current form, PMP models are not suitable for determining process lethality calculations. To make these calculations, it is necessary to know the z value and T_{ref} temperature, and to calculate F values over a range of changing temperatures.

Cooling Models

The cooling/growth models for *C. botulinum* and *C. perfringens* are one of the most used models by the food industry. The USDA Food Safety and Inspection Service "Compliance Guidelines for Cooling Heat-Treated Meat and Poultry Products" states that during cooling, the product's maximum internal temperature should not remain between 130°F and 80°F for more than 1.5 hours nor between 80°F and 40°F for more than 5 hours (http://www.fsis.usda.gov/OA/fr/95033F-b.htm). When cooling for a product deviates from the compliance guidelines, the cooling model can be used to estimate if the deviation would result in unacceptable growth of pathogens to determine if the product can be salvaged.

Miscellaneous Models

These include gamma-irradiation models for *S. typhimurium*, *E. coli* O157:H7, and native microflora in meats, time-to-toxigenesis model for *C. botulinum* on fish, and time-to-turbidity models for *C. botulinum*.

Choosing a PMP Model

Models are developed from environmental conditions that are relevant to the objectives of a specific study. For example, the model may have been developed from data generated in a microbiological broth that is intended to simulate a food product or a group of food products, in a real-food product, or in a model food system. In each case, predictions from the model are known to be accurate to the specific food or system, and conditions under which that the model was developed. Therefore, it is critical to learn the specific conditions under which a model was produced to apply the predictions to other types of foods and conditions.

The associated references that state the specific system and the conditions under which a model was developed are included in PMP.

In PMP, there are multiple forms of a model, such as aerobic and anaerobic storage conditions. In this case, choose the model that is closest to your product. For example, choose an anaerobic model if your product is vacuum-packaged. Choose an aerobic model if you wish to understand how the bacteria will react when the package is opened and exposed to oxygen. In general, if bacteria can grow under either aerobic or anaerobic conditions, growth is typically faster under aerobic conditions.

In many cases, you will not find a model that matches your food product formulation. In this case, it is better to choose a model that will provide more liberal growth estimation or conservative inactivation prediction. For example, culture media (broth) models typically predict shorter generation times (or higher growth rates) than those observed in food containing other microorganisms and additives. Similarly, a sterile raw food model will normally predict shorter generation times (or higher growth rates) than a nonsterile raw food.

Operating the PMP Program

The model predictions were developed for a specific range of environmental conditions. For example, the Aerobic *A. hydrophila* in Culture Media model was developed over a temperature range of 41°F (5°C) to 107.6°F (42°C). The accuracy of predictions made inside (interpolation) this range is known. However, the accuracy of predictions made outside (extrapolation) of this range (e.g., 150°F) is not known. The software does not permit values outside this range of temperature to be entered.

Model Interpretation

Without experience in the use of models, it is difficult to know if the model that you use will over- or under-predict bacterial growth or inactivation when applied to another food matrix. As such, it is best to use models to understand potential trends in bacterial behavior as the environmental conditions change. Only through validation studies (e.g., inoculated pack studies) would you be able to have confidence in model interpretation for your specific food of interest.

Depending on the pathogen, spoilage flora (e.g., bacteria and fungi) can markedly inhibit the growth of pathogenic bacteria. This is especially apparent at refrigeration temperatures where the growth rates of psychotropic (cold-liking) spoilage organisms may be greater than that of the pathogen.

Therefore, in these situations, the MPD (level at stationary phase) of a pathogen may be 3–5 $\log_{10}$ levels less than that observed in a pure culture. Also, the growth rate may be inhibited. Therefore, in general, generation times will be shorter, and growth rates and maximum population densities will be higher in sterile culture systems compared to systems containing spoilage flora. Again, only through validation studies (e.g., inoculated pack studies) would you be able to have confidence in model interpretation for your food of interest.

Using a Static Temperature PMP Model for Changing Conditions

Currently, there are dynamic temperature models for *C. botulinum* and *C. perfringens*. For all other models, predictions are made for static temperatures. To use a static model for situations where temperature changes over time, the growth at each of the temperatures needs to be calculated, and then these individual growth calculations are added to determine the total predicted growth over the entire time–temperature range. The following example uses the PMP *L. monocytogenes* aerobic model to predict the total growth of *L. monocytogenes* in cooked chicken, which has the following time–temperature profile:

0.0 hour 98.6°F
0.5 hour 71.2°F
1.5 hours 63.4°F
3.5 hours 50.1°F

Procedure

First, for fail-safe predictions, assume that the product was at 98.6°F for 0.5 hour, at 71.2°F for 1 hour, and at 63.4°F for 2 hours. Set the environmental conditions that are close to the cooked chicken product such as pH = 6.5, NaCl = 1.0%, 0 ppm nitrite.

- Set the initial level of *L. monocytogenes* at 3.0 $\log_{10}$ CFU/g (Fig. 7.2).
- Set the temperature to 98.6°F (37°C)
- Click the box "calculate growth data."
- Click the "no lag" box (for fail-safe predictions).
- Click "show table" under "display format."
- Average the counts for 0.4 (3.64 log[CFU/mL]) and 0.6 hour (3.76 log[CFU/mL]) to obtain the counts increase after 0.5 hour at 98.6°F (30°C) of 3.70.

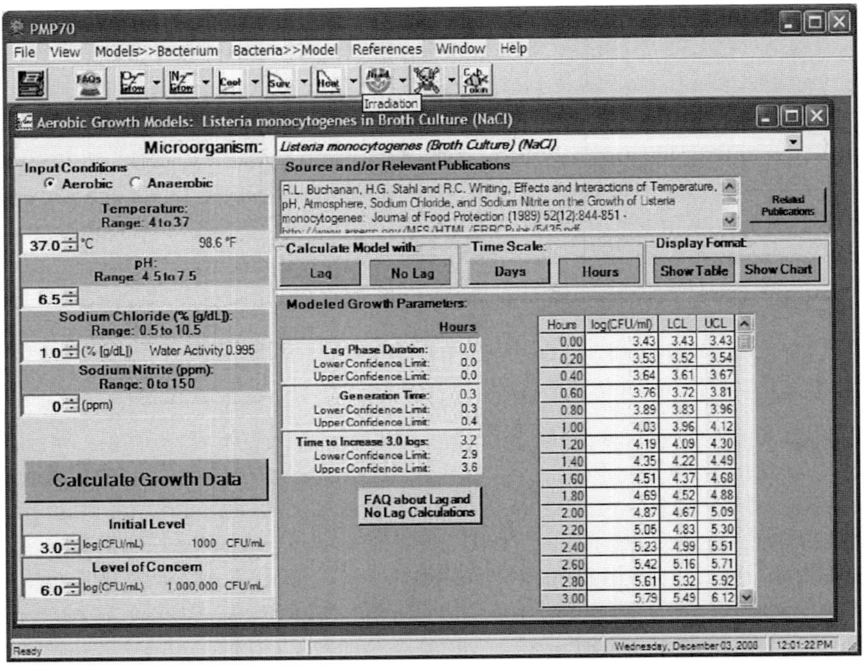

Figure 7.2. Input and output of PMP *Listeria monocytogenes* model.

- Subtract 3.0 (starting level) from 3.7 and the resulted 0.70 log CFU/mL is possible growth of *L. monocytogenes* at 98.6°F (30°C) for 0.5 hour.
- Repeat the same procedure for other temperature/time predictions. The resulting increases were 0.63 log at 71.2°F (21.8°C) for 1 hour, and 0.67 $\log_{10}$ at 63°F (17.2°C) for 2 hours.
- The predicted total increase in *L. monocytogenes* count in the chicken product under the temperature profile is $0.70 + 0.63 + 0.67 = 2.0$ log.
- As with all PMP models, you need to validate the predictions for food types and formulations that are different from the model.

Using Models in HACCP Plans

Models are only valid for the conditions used to produce the model. The reference(s) found in the PMP "Source and/or Related Publications" presents an explanation of the methodologies used to produce the model. Therefore, if the conditions (e.g., food formulation) used to produce the PMP model do not match your food system, then you must validate the model for your specific application. Validation normally involves independent laboratory studies where your product is inoculated with a specific bacteria and then you record the levels of growth or inactivation. These data can then be compared with model predictions to see if they are within the predicted 95% confidence intervals, and/or if they are fail-safe. If they do not match, then the model is not valid (safe) for your application and should not be used in HACCP plans to make safety decisions. Assuming sufficient experimental data have been collected, your data may instead be used to develop a new model that would be valid for your food product.

Summary

The PMIP is a comprehensive web portal developed by the Microbial Food Safety Research Unit at the ERRC of USDA-Agriculture Research Service. It is a one-stop resource designed to help small and very small processing companies to make better food production and management decisions by providing access to predictive models, regulatory policies/guidelines, and other information of relevance to the safety, quality, and wholesomeness of foods. This searchable database allows users to find information that could be used to develop plans for hazard analysis and critical control point, a system of programs for promoting food safety. The key feature of this web portal is a tutorial section with instructions on using and interpreting predictive models. PMIP currently offers information on research, regulations, and resources related to the fate of *L. monocytogenes* in ready-to-eat foods. The portal directly links users to USDA-ARS-PMP and ComBase. The PMP presently contains some 42 models, of which 65% are directly in foods and 35% are broth models, and includes both static and dynamic temperature models. These models allow users to predict food formulation, as well as processing and handling conditions, to control the growth, survival, and death of various bacterial foodborne pathogens. The PMP has become a premier multilingual, international modeling tool that is also used by government agencies and food processing companies in the management of food safety systems and is downloaded more than 8000 times each year in over 35 countries. Once downloaded, user-friendly features allow the client to easily input food-relevant criteria and then to receive predictions about how pathogenic bacteria react to specific food environments. ComBase, a relational database of predictive microbiology information, is a major international initiative to coordinate the collection and dissemination of food microbiology data. It contains >40,000 data sets that describe the growth, survival, and inactivation of bacteria under diverse environments relevant to food processing operations.

References

Albert, I., and Mafartb, P. 2005. A modified Weibull model for bacterial inactivation. *International Journal of Food Microbiology* 100:197–211.

Chen, H. 2007. Use of linear, Weibull, and log-logistic functions to model pressure inactivation of seven foodborne pathogens in milk. *Food Microbiology* 24:197–204.

Jay, J.M. 2000. *Modern Food Microbiology*, 6th edn. Aspen Publishers, Gaithersburg, MD.

Le Marc, Y., Pin, C., and Baranyi, J. 2005. Methods to determine the growth domain in a multidimensional environmental space. *International Journal of Food Microbiology* 100:3–12.

Tamplin, M.L., Paoli, G., Marmer, B.S., and Phillips, J. 2005. Models of the behavior of *Escherichia coli* O157:H7 in raw sterile ground beef stored at 5 to 46°C. *International Journal of Food Microbiology* 100:335–344.

Virto, R., Sanz, D., Alvarez, I., Condon, S., and Raso, J. 2006. Application of the Weibull model to describe inactivation of *Listeria monocytogenes* and *Escherichia coli* by citric and lactic acid at different temperatures. *Journal of the Science of Food and Agriculture* 86:865–870.

CHAPTER 8

Supporting Documentation Materials for HACCP Decisions

Mary Kay Folk, The Ohio State University

Introduction

The supporting documentation materials for HACCP decisions were written with the intention of assisting meat and poultry processors in the scientific justification of their hazard analysis and critical control point (HACCP) decisions. The manual was compiled from extensive literature searches and written in a manner for all processors and inspectors to understand and use. The information is carefully summarized from each scientific source and categorized by HACCP process. This information can be used to scientifically justify decisions made related to good manufacturing practices (GMPs), sanitation standard operating procedures (SSOPs), critical limits, corrective actions, or any other HACCP decisions applicable to a specific process.

The documentation manual is meant to be a starting place for processors to find scientific information as part of the validation procedure of HACCP. The hazard analysis portion of the HACCP process requires processors to understand where a hazard is likely to occur and how it is likely to become a hazard. Understanding this "where" and "how" can be a problem for processors unless they are sufficiently trained in possible causes of hazards.

Background

Early in the HACCP implementation period, USDA's Food Safety Inspection Service (FSIS) published *Generic HACCP Models* to give establishments and examples of HACCP plans, to assist them in the development of their plans. These generic models were written for the categories: pork slaughter; beef slaughter; poultry slaughter; raw, ground product; raw, not ground product; not heat-treated, shelf-stable product; heat-treated, shelf stable; fully cooked, not shelf-stable; heat-treated, not fully cooked, not shelf-stable; products with secondary inhibitors, not shelf-stable, mechanically separated (species)/mechanically deboned poultry; thermally processed, commercially sterile; and irradiated products. They were meant to be a reference source for each of the categories with an example of what an HACCP plan might include. However, they were not meant to be all encompassing and they were not meant to be an actual plan for any establishment. The models do give processors some scientific resources for those specific examples but, again as examples, they only addressed selected process steps.

In the original implementation of HACCP, Principle number 6—"establish procedures to verify that the HACCP system is working correctly" is laid out as an integral step to the safety of the system (9CFR 417.4). Supporting documentation materials, such as published scientific literature, would be important in the initial validation and reassessment components of the verification principle.

In the push to implement HACCP in all meat and poultry processing plants, this validation step was somewhat overlooked by FSIS. However, once the very small establishments had implemented HACCP in January of 2000, the next FSIS objective was to revisit the HACCP plans and the required validation processes. Validation was initially defined by the National Advisory Committee on Microbiological Criteria for Foods (1998) as the part of the verification principle of HACCP that involves "collecting and evaluating scientific and technical information to determine whether the HACCP plan, when properly implemented, will effectively control the hazards." Supporting documentation, for validation, can include scientific publications, regulatory documents, company experimental data, or pathogen test results, etc. (Scott, 2005).

Scientific publications are the most common form of supporting documentation and refereed publications from scientific journals are the most acceptable form of scientific publications. However, some reference books that are based on technically sound data can also be used as supporting documentation.

FSIS assumed that large companies would have the resources and personnel to take on this validation task, but the small and very small companies may lack the technical resources for both locating the scientific information and correctly applying it to their situation. Therefore, the Ohio State University's supporting documentation materials were written with small and especially the very small establishments in mind. However, as the manual was made available, it was noted that even the largest of processors found this new resource helpful.

Purpose

The manual is intended to give meat and poultry processors access to a scientific resource and put it into a format that can be easily matched to the parameters of their own processes. Building on those selected steps covered in the FSIS examples, the supporting documentation manual addresses process steps that have been covered by published research. This includes circumstances that are not typical in a processing environment, however, could occur given a mistake is made. Some of the information in the material addresses formulation, not a specific step, but impacts how other steps may be handled.

Because of the wide range of information in this compilation, the scope of HACCP decisions that are addressed is vast. For instance, one establishment may use information concerning minimum growth temperature of *Salmonella* on pork as a justification for a critical limit, where another company may use the same information to justify a good manufacturing procedure (GMP) that assures the safety of a product.

Organization

The information in the documentation manual is arranged according to the ten process categories defined by FSIS: red meat slaughter; poultry slaughter; raw product—not ground; raw product—ground; fully cooked, not shelf-stable; heat-treated, shelf stable; heat-treated, not fully cooked, not shelf-stable; not heat-treated, shelf stable; secondary inhibitors, not shelf-stable; and thermally processed—commercially sterile. The information gleaned from the references mentioned earlier was organized by the process step within each of the process categories. Some research covers several products that fit into different processing categories. In this case, the information was added to all of the appropriate categories at

Table 8.1. Overview of information included in supporting documentation materials

Process Step	*Potential Hazards*	*Process Parameters*	*Decision Criteria*	*Scientific Documentation*
This column indicates the point or step of each process flow, in which scientific or regulatory documentation is available	This column identifies the potential hazards that have been addressed in published scientific literature, for each process step	This column describes the conditions used in the research that is described in various scientific publications	This column describes the results of the research, or the regulatory requirements	This column describes the actual source of the information, described in the three columns to the left. Where available, a web site is given to allow the internet access to publications

the appropriate processing step. Other research covers several steps related to the same type of product. This research would be entered into the same process category under each applicable processing step. Additionally, there are separate chapters for physical hazards and irradiation because both apply to all or most of the categories, and this repetition was avoided. An overview of the information found in each column in the supporting documentation is given in Table 8.1. The following will describe the process more thoroughly.

Once the processing category and steps were defined from a scientific publication, the particular hazard was identified and entered into the second column of Table 8.1. As mentioned earlier, the biohazards deal only with pathogenic organisms. In a few processing steps where pathogen research was not available, articles that address only spoilage bacteria were included. The "potential hazard" information also includes the method in which the hazard is addressed in the research. These include pathogen survival (e.g., cooking), pathogen growth (e.g., storage temperature vs. time), and pathogen contamination.

The next piece of information taken from each document was the test parameters. Priority was given to scientific publications that use meat products as the test medium; however, some articles were included that used broth or liquid growth media. This section then addressed the specifics about those products that the article provided. Some examples are species, pH level, salt content, nitrite content, and spices. The parameters also include the conditions that the test media, whether that be meat or liquid, were exposed to during the study. Temperature, oxygen level, packaging material, and pressure and duration of sprays are just some examples of conditions included here. Some of the parameters are not and should not be found in daily processing in the USA. Many of the extreme situations, though, may be found when a process goes out of control (e.g., the ovens or the coolers fail, product is not stored in the correct place). There are many instances in which product is not treated in a manner that is known to be safe, but this research can help to elucidate if and at what point it became unsafe. Some of the research reported in the supporting documentation manual tested additives above legal limits for use in the USA. Others were not allowed at any level at the time of publication. In both cases, this research was included to show the effectiveness or ineffectiveness of various intervention strategies, even when they are not allowed currently. Regulations may change to allow some interventions, but the research may show that levels higher than that allowed are not beneficial anyway.

Finally, the last piece of information added into the document was the results of each study. This column answers the question, "How did the pathogens respond?" This column reports logarithmic growth (log 10) or destruction of the pathogenic bacteria in question. The FSIS regulations are written using log numbers as requirements or limits to the degree of pathogen destruction or growth that must occur during various processing steps.

All scientific publications in the manual are referenced in the same row with the information gleaned from that document. This not only gives credit to the appropriate scientists, but also this allows the user to find the original article. A copy of the full article is required by FSIS to accompany an HACCP plan when it is used to validate a decision.

One example of a reference in the supporting documentation material is the 1978 *Journal of Food Protection* article written by Goodfellow and Brown, from which the FSIS Appendix A is based. Table 8.2 shows a portion of the information drawn from that resource. Where available, direct links to publications are provided in the last column of this document.

Table 8.2. Example reference from supporting documentation material for HACCP decisions

Process	*Potential Hazards*	*Process Parameters*	*Decision Criteria*	*Scientific Documentation*
Cooking	B—*Salmonella* survival during cooking process	Dry roasting of large beef roasts at oven temperatures of 250°F (121°C) or 275°F (135°C)	*Salmonella* will be destroyed (7-log reduction) if roasts (16–18 lb) are dry roasted to these specifications:	Goodfellow, S.J., and Brown, W.L. 1978. Fate of *Salmonella* inoculated into beef for cooking. *Journal of Food Protection* 41(8):598–605
			250°F (121°C) oven, internal temperature of at least 130°F (54.4°C)	
			275°F (135°C) oven, internal temperature of at least 125°F (51.6°C)	
		Dry roasting small (less than 10 lb) beef roasts in oven temperatures of 275°F (135°C) or less	*Salmonella* are not fully destroyed when dry roasting beef roasts of less than 10 lb in an oven at 275°F (135°C), or less, when heated to an internal temperature of 135°F (57.2°C); however, there was a 5-log reduction	

Resources

The supporting documentation manual was compiled from extensive literature searches of refereed journal articles, reference books, USDA-FSIS, and Food and Drug Administration (FDA) regulations. Journal articles that were used focused on processing steps or other treatments that affect or hypothesize to affect pathogen growth cycles and logarithmic concentration of those pathogens. Much research has been published with indicator and spoilage bacteria. However, at the outset of this project the decision was made to strive for using only information applying to pathogens. In a few cases, spoilage bacteria research was included when appropriate research at a specific processing step using pathogens was not found. Another focus of the materials used in this compilation was research performed using meat as the testing medium. A number of articles were included that did not use meat when research could not be found that specifically pertain to meat products.

References from the above-mentioned FSIS models were used to ensure that all publications that FSIS had recognized were included. Sources from those articles were the next vein of references explored. Databases of journal articles were searched for specific pathogens, meat types, and processes. Textbooks provided basic information, but when possible the original journal article was found so that the information reported could be in uniform structure. Throughout the compilation of this scientific information, pertinent federal regulations from the Code of Federal Regulations and the FDA Food Code were added at the appropriate processing steps.

Many research articles were included in spite of the reported findings. In some cases, results did not agree with UDSA-FSIS regulations and guidelines that have been established. This type of research is important to show that there is variation to the methods that produce a safe product. Other research compared parameters that were not within FSIS regulatory limits. Some of this research was done outside of the USA where the regulatory limits are different. Other research showed that even at extreme levels the treatments were ineffective. These results were included in this manual because it is important to show that the research has been done and demonstrated the possible effects, partially, so that others will not spend time or resources repeating these treatment comparisons. Further research was included that tested conditions that are not reasonable for normal processing. These results may not be applicable to the average processing operation, but they demonstrate the extent of hazards when the process is not under control, corrective actions must be taken, and decisions associated with those actions need to be justified.

Distribution and Access

In 2002, USDA-FSIS printed 10,000 copies of the supporting documentation material for HACCP decisions. These were distributed to all federally inspected small and very small meat processors and many of the state inspection programs. Copies of this document have also been available to larger establishments, by requesting them from FSIS or the Ohio State University Meat Science program.

The documentation manual has always been considered a living document by the authors. The body of research available is continually growing. Therefore, the supporting documentation manual is continually being updated. The original printing was quickly outdated. Knowing that this would be an ongoing problem, the manual was made available in electronic form on the internet even before that first printing. The link can be found on the FSIS science resources web site. That link takes the user to the Ohio State University Meat Science HACCP Support web page. The direct link to that page is:

www.ag.ohio-state.edu/˜meatsci/HACCPsupport

This page is updated as new scientific articles are found and added to the documentation manual.

A second printing of this supporting documentation materials was done in 2007, and hard copies can be requested from FSIS through their web site at: http://www.fsis.usda.gov/Science/HACCP_Resources_Order_Form/index.asp.

Conclusion

The supporting documentation material for HACCP decisions was written with the expectation to provide small and very small meat processors with a resource that would aid in the scientific validation of their HACCP decisions. The document has exceeded the authors' expectations in the audience that found the information useful. From the outset the manual has been considered a living document. Scientists will continue to build upon the current knowledge base and new technologies will become available to the average processor. The continually changing nature of the field demands that we meet the challenges in new and innovative manners. We hope that this document will continue to be a resource for those looking to employ fresh and practical information.

References

9CFR 417.4. Available at: http://www.access.gpo.gov/nara/cfr/waisidx_06/9cfrv2_06.html#301.

Food and Drug Administration. Food Code. Available at: http://www.cfsan.fda.gov/~dms/fc05-toc.html.

Goodfellow, S.J., and Brown, W.L. 1978. Fate of *Salmonella* inoculated into beef for cooking. *Journal of Food Protection* 41(8):598–605.

NACMCF (National Advisory Committee on Microbiological Criteria for Foods). 1998. Hazard analysis and critical control point principles and application guidelines. *Journal of Food Protection* 61(9):1246–1259.

Scott, V.N. 2005. How does industry validate elements of HACCP plans? *Food Control* 16(6):497–503.

CHAPTER 9

The Ten Principles of Sanitary Design for Ready-to-Eat Processing Equipment

David Kramer, Sara Lee Corporation

In 2000, the American Meat Institute (AMI) recognized the need to improve sanitary design as a key initiative toward continuously improving food safety and established the first task group to develop the principles of sanitary design for equipment. The AMI Task Group assigned to develop these principles was primarily focused on postlethality contamination of ready-to-eat (RTE) products. That is the reason the principles were titled for RTE equipment. However, it is safe to say that everyone involved in the work would agree that these principles are universal and apply to all equipment in your facility whether it is in the raw or RTE areas. A key objective of sanitary design is to make sure the initial microload going into your lethality process is no higher than the load assumed in the lethality calculations.

This chapter discusses the AMI principles of sanitary design for equipment. The next chapter discusses the AMI principles of sanitary design for facilities.

Principle 1: Cleanable to a Microbiological Level

Principle 1 requires that food processing equipment must be constructed and maintainable to ensure that the equipment can be effectively and efficiently cleaned and sanitized over the life of the equipment. This includes both product and nonproduct contact surfaces of the equipment. The goal is to prevent bacterial ingress, survival, growth, and reproduction. The key to achieving this goal is to understand the life cycle of a microorganism and to begin thinking about the scale of the enemy we are trying to eliminate. A crack, crevice, or pit on a piece of equipment may look incredibly small to our eyes but it would appear like the Grand Canyon to a microorganism. Your equipment designer needs to understand this and constantly confront the issue of how are we going to clean this piece of equipment to a microbiological level every day. Critical to this are the adverbs effectively and efficiently used above. If the design requires a high-level maintenance technician and 10 hours of time to disassemble, clean, sanitize, and reassemble, it is highly unlikely that this will occur frequently enough to achieve your goal. Food processing equipment must be designed to be effectively and efficiently cleaned within the economic time and cost constraints required for profitable operation.

Principle 2: Equipment Must Be Made of Compatible Materials

Principle 2 requires that the materials of construction for food processing equipment must be completely compatible with the product, environment, cleaning and sanitizing chemicals, and the methods of cleaning and sanitation. There are four criteria that the proper materials of construction must meet in order to achieve compatibility. First, they must be inert. This means that they cannot chemically react with your product, other materials that may be in contact or processing agents used in your process. If your material selection is not inert, the by-products of the chemical reaction will become adulterants in your product, which can affect flavor, bind, and shelf life. In general, you want to avoid using soft metals such as copper, bronze, and aluminum whenever possible. Second, they must be corrosion resistant. This is, in fact, a subset of being inert. It is mentioned separately because corrosion creates surface defects which can become likely niches and harborages for microorganisms. These small surface defects are beyond the reach of our normal cleaning procedures. Their geometry prevents the bristle of your brush or your green pad from getting down into them to provide the necessary mechanical action to clean them. Third, they must be nonporous. I like to think of porous surfaces as microscopic sponges. If these surfaces can retain water, it is highly likely that they will also become niches and harborages for microorganisms which are uncleanable with our normal cleaning processes. Fourth, they must be nonabsorbent. Unsealed, fabric reinforced belting is the most often cited condition whereby the absorbent properties of the fabric actually draws moisture into the fabric and can create a perfect niche for microorganisms to grow. Less obvious are some plastics, such as nylon, which can readily absorb moisture and create difficult to clean niches. You just need to check the material specifications carefully to make sure your material selection is not creating another problem for you.

Principle 3: All Areas of the Equipment Must Be Accessible for Inspection, Maintenance, Cleaning, and Sanitizing

Principle 3 can be summed up by simply stating that if you cannot see it and cannot reach it—you cannot clean it. Likewise, if you cannot see it and cannot reach it, you will not be able to effectively inspect it and sample it. This means you will not be able validate your sanitation procedures which also means that you are relying on luck as the foundation of your sanitation program. This is totally unacceptable. Your equipment needs to be designed to provide complete, safe access to every square inch of surface area. This requires the design to include the necessary access doors and panels that meet the sanitation requirements for accessibility while also maintaining the Occupational Safety & Health Administration (OSHA) required guarding requirements during operation. It requires your design to be able to be fully disassembled so that all surface areas can be effectively cleaned and sanitized on a daily basis. Needless to say, this must be capable of being done in an efficient and economical manner given the constraints your plant schedule and manpower places on your sanitation process. The key to this is a design that allows disassembly with simple tools that can be easily handled by your sanitation staff.

Principle 4: Equipment Is Designed to Prevent Product or Liquid Collection

Principle 4 requires that equipment is designed to be self-draining to assure that food product, water, or other liquids do not accumulate, pool, or condense on the equipment, particularly, in product contact zones of the equipment. The reasoning is simple. Food + water + a protected environment inside a piece of equipment + 16-hour run times is like Daytona Beach at spring break for microorganisms. If your equipment design does not adhere to Principle 4, you may well be left with no choice other than to perform mid-shift disassembly and cleaning to control the situation. I will guarantee you that this is not an efficient or economical solution. Additionally, you may run the risk of product adulteration from your cleaning chemicals since those areas that do not self-drain are likely to hold detergents or sanitizing compounds which may find their way into your product once the start button is pushed in the morning. The unfortunate issue with equipment that does not adhere to Principle 4 is that the areas that do hold product or liquid may accumulate product for far longer than a single production day since they are normally not very accessible. This only exacerbates the food safety problem and can lead to the development of biofilms that may be very difficult to remove.

Principle 5: Hollow Areas of Equipment or Components Are Hermetically Sealed

Hollow areas of equipment such as frames and rollers must be eliminated where possible or permanently sealed. Caulking is not an acceptable solution. Bolts, studs, mounting plates, brackets, junction boxes, name plates, end caps, sleeves, and other such items must be continuously welded to the surface of the equipment and not attached via drilled and tapped holes or rivets. Hollow areas tend to provide the ultimate niche because they normally cannot be accessed for proper sanitation without physically cutting your equipment apart. Once water gets into a hollow area microbial growth will start. Since the areas are inaccessible, your normal inspection and sampling program will not detect them until the liquid begins to leak out into your production area. Then you will begin to pick up the trail with your environmental sampling program but finding the actual source may take weeks unless you have an experienced equipment sleuth on your staff. When faced with this problem, I like to assemble a team from maintenance and quality assurance and spend a weekend drilling holes into the lowest points on the hollow framework of the equipment to see what runs out and get a sample for laboratory analysis. The holes are then plug welded, samples analyzed, and an action plan developed to address any issues uncovered.

Hollow rollers are a particularly insidious sanitary design problem. They provide an ideal niche for microorganisms to proliferate. Most hollow rollers are not hermetically sealed and those that are rarely remain so for the life of the equipment. This allows water to enter the hollow cavity during operation or sanitation. The water carries enough nutrients and microorganisms into the cavity to begin the growth and proliferation of the microorganisms. The tumbling action from the turning hollow roller acts much the same as a gardener's compost tumbler by aerating the mixture and uniformly distributing the nutrients. For the gardener this results in rapid composting due to increased microbiological activity. For the food processor it results in higher numbers of microorganisms and a continuing source of trouble. Once the microorganisms are established inside the hollow roller, the cycle of recontamination begins as soon as the hollow roller starts to turn. Unfortunately, many sanitation workers and plant personnel remain unaware of the inherent risks from hollow rollers. So the cycle of cleaning, sanitizing, inspecting, operating, and recontamination will continue until the hollow rollers are eliminated. Hollow rollers must be eliminated from our designs wherever possible. I have found very few instances where easy engineering solutions did not exist to replace hollow rollers. Where hollow rollers need to be used due to size and weight considerations they must be hermetically sealed and should contain a brightly colored nontoxic food grade dye to facilitate discovery of the loss of the hermetic seal. I have one final caution with respect to hollow rollers. Hollow rollers have a nasty habit of reappearing. This happens because equipment manufacturers do not always change their parts manuals to reflect a design change for equipment already in the field or notify their customers of their change to solid or hermetically sealed rollers. So, when your mechanic orders new rollers he is often receiving the originally specified hollow rollers instead of the improved design. Equipment manufacturers and food processors need to take the same steps to prevent the reinstallation of hollow rollers—change your parts manuals and purge the parts room of hollow rollers.

Principle 6: No Niches

All parts of your equipment must be free of niches such as pits, cracks, corrosion, recesses, open seams, gaps, lap seams, protruding ledges, inside threads, rivets, and dead ends. All welds must be continuous and fully penetrated. These conditions create uncleanable areas where microorganisms find protected growth niches that result in continuous recontamination of your equipment surfaces until the niche is eliminated. Sanitizers are not a silver bullet that will resolve niche problems. The niches must be eliminated through repair and/or redesign that facilitates proper sanitation. Over time, niches will develop in the best sanitarily designed equipment due to metal fatigue, maintenance activity, and normal wear and tear on the equipment. You must have a good equipment audit program in place to control niche development and correct problem areas as they arise.

Principle 7: Equipment Must Be Designed for Sanitary Operational Performance

This principle moves us from the fabrication shop to the operational world of your production facility. Designs that may look good on paper and in the vendor's shop may not perform as expected during operation. Normal operational conditions must not contribute to unsanitary conditions or the harborage and growth of bacteria. This can readily happen if the equipment accumulates fine particles of meat during the day. These fines sit in a warmer environment due to the mechanical heat generated by the operating equipment and can contribute to accelerated bacterial growth over a 16-hour production day. Often this situation develops in areas of the equipment that require OSHA guarding during operation for personnel safety. Guards on cutting equipment such as slicers and saws come to mind as good examples. Another good place to look for trouble is on conveyors equipped with belt scrapers. The best way to find these problem areas is to perform a periodic inspection of new equipment in its first few days of operation. Obviously, this will require that you follow all specified safety precautions such as lock out/tag out. Often these issues are very difficult to correct because the best food safety answer does not meet the personnel safety requirements. Nonetheless, you will have to either work with the equipment manufacturer to redesign a solution that meets both requirements or you will have to put administrative procedures in place that control the food safety hazard.

Principle 8: Hygienic Design of Maintenance Enclosures

Maintenance enclosures such as electrical control panels, chain guards, belt guards, gear enclosures, junction boxes, pneumatic/hydraulic enclosures, and associated human–machine interfaces such as pushbuttons, valve handles, switches, touch screens, and latches must be designed and constructed and maintainable to ensure food product, water or product liquid does not penetrate into, or accumulate in or on the enclosure and interface. Principle 8 addresses an often unnoticed and unidentified danger zone in your facility. For safety reasons, many of the enclosures listed above are off limits to your production, sanitation, and quality assurance personnel, so unsanitary conditions inside of these enclosures is unlikely to be reported unless you have thoroughly trained your maintenance technicians to recognize the hazard. Once again, these are often areas that are warmer than the surrounding environment due to the heat generated by the mechanical or electrical components and therefore provide a nice incubation environment for microorganisms to thrive. A typical scenario that you will find is a production supervisor places a trouble call to maintenance. They dispatch a technician to troubleshoot the problem. The technician opens up a maintenance enclosure, reaches in to test or adjust a component then reaches up to push the start button on the piece of equipment to see if he has fixed the problem. If the maintenance enclosure was contaminated he has now contaminated the start pushbutton. The technician, having fixed the problem, leaves and your operator reaches up and hits the start button. Now, your operator's hand is contaminated and you know the rest of the story. Maintenance enclosures and associated interface devices need to be periodically sampled to determine the proper frequency for cleaning. Problem enclosures need to be closely examined to determine how the contamination is entering the enclosure and remedial action needs to be taken to resolve the issue.

Another problem area with these enclosures concerns their basic construction. For many years the standard enclosures on the market had flat tops. Obviously, a flat top is not free-draining (Principle 4) and the flat top often serves as a shelf for product, hand tools, or supplies. This creates a whole other list of contamination vector possibilities that you will have to deal with. Today, there are very nice sloped top enclosures available for the good manufacturing practices market that eliminates the problem. These should be specified with your equipment vendor.

Principle 9: Hygienic Compatibility with Other Plant Systems

Equipment as designed built and shipped is worthless until it is installed with all of the required utility connections, exhaust ducts, drains etc. Yet, the impact of these connections is often ignored in the design process.

I have seen designs where the physical connection point for the utility connections actually blocked the accessibility to a critical removable panel on the piece of equipment once the pipe and conduit were installed. I have seen other designs where the connection points dictated that the utilities would be over exposed product areas. These are design oversights that can and should be remedied by the equipment vendor. A larger issue for the industry is our own installation practices. You cannot rely on the electricians, pipefitters, or plumbers to make these utility routing decisions in the field. Your facility designers need to plan and specify the exact routing in order to avoid food safety conflicts. Make sure they will not interfere with your sanitation process and remember that all surfaces in your production areas must be cleanable to a microbiological level. Specify accordingly.

Principle 10: Validated Cleaning and Sanitizing Procedures

You are required by the USDA to have standard sanitation operating procedures as part of your overall HACCP plan. These procedures must be clearly written, designed, and proven to be effective and efficient. Chemicals that are recommended for cleaning and sanitizing each piece of equipment must be compatible with the equipment and the manufacturing environment. Normally, this has been done as an exercise after the equipment has been purchased and installed. This process leaves the equipment designers in the dark. They do not know if you plan to use city water pressure or 1,000 psig system to clean their equipment; yet, I think we would all agree that you would design the equipment a bit differently for each scenario. Eventually, I would like to see equipment manufacturers bear the responsibility for specifying sanitation procedures for their equipment. This will require them to become more knowledgeable about the sanitation process which will result in better sanitary designs. Ultimately, you will always be responsible for executing the sanitation protocol and validating its effectiveness through sampling.

In the end, good sanitary design needs to be combined with good sanitation practices in order to deliver the results you need. Cleaning is a three-step process. First, you have to be able to find the soils you want to clean. The AMI Principles of Sanitary Design place a lot of emphasis on accessibility. If your sanitation crew cannot easily reach or see the surfaces to be cleaned, they will not find the soils. Second, you have to be able to detach the soils from the surfaces. Surface finish, material selection, water temperature, and access to allow for the application of sanitation chemicals and mechanical force are critical for lifting the soils from the surfaces. Knowledge of the specific soils you are dealing with is a prerequisite for optimizing the material selection, surface finish, and sanitation chemical selection. Third, you have to be able to rinse the soils down the drain and out of your facility. Steps one and two are wasted efforts if you leave the soils in your facility. Here, the concept of self-draining surfaces is critical. Properly sloped floors and self-draining equipment and facility surfaces enhance your ability to achieve this task.

Sanitary design is not a destination. It is a journey of continuous improvement driven by observation, sampling, evaluation of new materials and techniques, and adjustment to the human, regulatory, and societal factors that influence your business and the markets you serve.

CHAPTER 10

Principles of Sanitary Design for Facilities

David Kramer, Sara Lee Corporation

Introduction to Sanitary Design

In 2003, the American Meat Institute (AMI) established a second task group to develop the principles of sanitary design for facilities. The application of these principles to your facility is critical to maintaining an environment that meets the microbial load assumptions used to design your thermal processes and to reduce the probability of postlethality contamination of ready-to-eat (RTE) products. The principles outline rigorous design concepts to improve overall cleanability and maintain your food safety standards.

Aesthetically clean surfaces are no longer a sufficient metric for assessing the sanitary condition of your facility. Surfaces must be microbiologically clean to ensure safe food products. The application of sanitary design principles provides two important hurdles for your facility's food safety program: barriers to the introduction and transfer of chemical, physical or microbial hazards and barriers to the growth of microorganisms or inadvertent transfer of allergens through improved cleanability of the designs.

The ultimate goal is a controllable environment for the production of your products. Sanitary design is an investment in your business. I like to think of it as an investment in brand name security insurance. What would your management be willing to pay today to avoid a future recall which could tarnish, or possibly, destroy your brand name? At issue is the fact that it is relatively easy to track the initial increased cost for improved sanitary design in our accounting systems.

It is very difficult to track the benefits through these same systems due to time lags and the multiplicity of ways in which poor sanitary design can manifest itself over time. Recalls, withdrawals, shelf-life issues, increased sanitation cost, missed production schedules due to recleans, and lost production due to downtime for repairs are all real costs that you may incur over time. I find it easier to sell the concept in terms of life-cycle costs over the expected life of the facility or piece of equipment. For example, if your facility produces 100 million pounds of product a year and you assume a 40-year facility life, you are looking at a life-cycle production of 4 billion pounds of product. If you are willing to invest only $0.001 per pound, this will give you 4 million dollars for improved sanitary design at a life-cycle cost of only one-tenth of a cent per pound. I believe that is a reasonable cost for protecting your brand and your consumers.

The 11 Principles of Sanitary Design for Facilities

The 11 principles of sanitary design for facilities are divided into three key concepts: create zones of control, keep it cold and control moisture, and facilitate the sanitation process.

Zones of Control

Principle 1: Distinct Hygienic Zones Established in the Facility

Principle 1 requires strict physical separations that reduce the likelihood of the transfer of hazards from one area of the plant, or from one process to another area of the plant or process, respectively. It also requires a facility design that facilitates the necessary storage and management of equipment, waste, and temporary clothing to reduce the likelihood of the transfer of hazards. In general, this means designing physical separations between raw and RTE areas of your facility. In slaughter operations, it may also include designing separations between preevisceration and postevisceration areas of your facility. The application of this principle to your facility requires you to familiarize your design team, construction team, and operation's team with your HACCP plan to make sure everyone understands the identified risks, control points, and critical control points. Then the team can decide on the actual boundaries for the required physical separation between different hygienic zones in your operation. At this stage of the process, it is usually best to focus on the boundary lines based on food safety requirements without regard to material or personnel movements. Principle 2 deals with personnel and material flows in your facility. Your final decision on the appropriate boundaries will be an iterative process based on the outcome of your Principle 2 analysis.

Principle 2: Control the Flow of Personnel and Materials to Reduce the Transfer of Hazards

Principle 2 requires you to develop a comprehensive process flow chart for the movement of all personnel and materials at your facility's site. Then you need to design controls for every point at which one of the flows crosses one of the boundary lines you established under Principle 1. This is probably the most difficult part of the sanitary design process because it introduces human psychology into the design process. We all have the natural tendency to take the path of least resistance. If your control points provide little physical restraint, they will be ignored. If they are too physically restrictive, your people will figure out shortcuts that reduce the effectiveness of the control point. From a designer's perspective, you have to focus on understanding how people will interact with your design. For instance, I once designed a hand wash, footwear sanitizing control point that just did not function to my expectations for one simple reason—I did not take into account that certain people need to enter the control point carrying items in their hands. I forgot to place a shelf near the sink where they could place the items while properly washing and sanitizing their hands. The only choice I had left for them was to place their items on the floor which defeated the purpose of the control point. Fortunately, my mistake was quickly and inexpensively remedied. Table 10.1 lists examples of some of the personnel and materials that require your attention when you begin to design your control points.

The industry trend is to provide separate raw and RTE welfare facilities for employees whose daily work is limited to each respective area. Table 10.2 lists examples of various control technologies and administrative

Table 10.1. What needs to be controlled

Personnel	*Materials*
Production workers	Raw materials
Supervision/staff	Packaging materials
Maintenance	Trash
Quality assurance	Rework
Medical	Inedible
Sanitation	Uniforms/laundry
Visitors	Maintenance parts
Suppliers	Sanitation chemicals
Regulatory	Quality assurance samples

procedures typically used to control the transfer of hazards associated with the flow of personnel and materials.

It is critical to understand that even the most rigorously designed physical control strategies ultimately rely on the administrative controls that you have in place. It only takes one instance of control failure to result in a food safety incident. Your control strategy must be audited regularly to verify its effectiveness. If the vending machine supplier's personnel are servicing the vending machines on the raw side of your facility prior to

Table 10.2. Design control elements

Access control systems
Doorway foamers
Handwash sinks
Footbaths
Boot scrubbers
Hand sanitizer dispensers
Color-coded uniforms and tools
Gowning vestibules
Captive footwear
Captive clothing
Separate RTE/raw welfare facilities
Dedicated RTE/raw maintenance personnel
Separate RTE/raw dry storage
Separate RTE/raw trash disposal

entering the RTE side of your facility you have a control strategy failure to deal with.

Keep It Cold and Control Moisture

Principle 3: Water Accumulation Is Controlled Inside the Facility

Water is the precursor for life as we know it. Wherever there is free moisture available you will find microorganisms that have adapted to the local conditions and thrive whether it is a mid-ocean vent, a geothermal hot spring, or a nuclear sludge pit. This is the reason scientists are so interested to learn if there is or ever was water on the lunar or Martian environments. So, controlling free moisture is a critical hurdle toward limiting the number and type of microorganisms in your facility. The drier your facility is, the easier it will be to control microorganisms. There are two reasons for this. The first and obvious reason is that water is absolutely essential for organisms to thrive. Our ancestors learned this many millennia ago when they noticed that the drying of food delayed spoilage. To this day drying is actively used as a food preservation technique. The second reason is that water often provides the vector by which microorganisms get transferred from one area of your facility to another. If your facility has pooling water or wet floors, I will guarantee that you are not achieving the environmental control that lets you sleep at night.

Water is also a critical raw material in our industry. So, our designs cannot eliminate water. You must control its flow and accumulation in your plant environment. The entire building system—floors, walls, ceilings, and supporting infrastructure—must be designed to prevent the accumulation of water. This is normally achieved by the proper selection of materials that do not absorb water and by the proper sloping of surfaces to facilitate free-draining of any water. Water discharges from plant systems such as cooling water from packaging machines must be piped directly to drains and not allowed to run across the floor to the nearest floor drain.

In practice, poor maintenance practices will cause a gradual deterioration even in a properly designed facility. Floors and drains require particular attention. Unsealed cracks or improperly repaired floor surfaces are prime growth niches for microorganisms. You need to remember that in a perfectly executed sanitation cycle you are relying on gravity to convey the water, detergent, soils, and microorganisms across the floor and down the drain. Cracks and old porous cast iron drains will retain moisture, nutrients, and microorganisms while providing a very comfortable harborage for growth.

Principle 4: Room Temperature and Humidity Are Controlled

Temperature and humidity are two growth factors for microorganisms that you can control through the proper design of refrigeration/heating, ventilation, and air conditioning (HVAC) systems in your facility. The first step toward achieving the proper design is establishing your target temperature and humidity control limits. Both temperature and humidity are highly process dependent and can vary substantially based on your particular processing requirements. In general, 39°F (4°C) is recognized as a good processing room temperature for controlling the growth rate of microorganisms while balancing the comfort of your workers. Properly designed airflow patterns with low airflow velocity are critical to achieving worker comfort at this temperature level. Humidity level should be controlled to achieve an air dew point low enough to prevent condensation in the room. For complete condensation control this means you have to identify the lowest surface temperature in the room and target an air dew point lower than that surface temperature. I like to design for an air dew point that is 5°F below the coldest surface in the room. Once you have established the design parameters for your rooms, your design professional will be able to select the proper equipment and control strategy to maintain these conditions.

Another important consideration is the inclusion of a clean-up purge system in your refrigeration/HVAC design in order to control the fog that occurs during sanitation when hot water hits the cold surfaces. Controlling the formation of fog during sanitation is a critical success factor for achieving a microbiologically clean room because your sanitation employees need to be able to clearly see the surfaces they are trying to clean. Essentially, a clean-up purge system delivers heated make-up air into your room while exhausting room air instead of recirculating it. The intent of these systems is to exhaust the water vapor laden air and replace it with drier outside air because it is impractical and cost-prohibitive to design a refrigeration/HVAC system with the coil capacity to remove the amount of water vapor introduced during the sanitation process. These systems use heated air because it will hold more water vapor than colder air and therefore enhance the removal of water vapor. It will also raise the temperature of room surfaces thereby raising the dew point at these surfaces, which limits the formation of condensation, which must be removed prior to the restart of production. If your startups are being delayed while your crew wipes condensation from overhead surfaces you will probably find a nice financial return with an investment in a clean-up purge system.

Principle 5: Room Airflow and Room Air Quality Are Controlled

Moving air is often another vector by which microorganisms may be transferred from one area to another. Airflow in a facility should always be designed to ensure that air flows from the highest level of hygiene areas to the lowest level of hygiene areas. In general, this means that air should flow from your RTE areas to your raw areas. While this sounds fairly straightforward, in practice it is devilishly difficult to achieve—particularly in older facilities. Air behaves a lot like water as it always seeks its own equilibrium. For every cubic foot of air that is exhausted an equivalent mass of air must enter to maintain this equilibrium. Unfortunately, many facilities are improperly designed or improperly modified so that the exhaust air volume is not matched by an equal amount of conditioned make-up air. This imbalance impedes the proper function of your exhaust systems due to higher local differential pressures in your facility. This may result in a change in airflow patterns within your facility and a reduction in the exhaust of heat, moisture, and particulates from your processes which may create a human health hazard as well as a food safety hazard. When this condition exists, air infiltrates into your building through open doors, cracks, or other building openings. Since this air infiltration is not conditioned by filtration, temperature control, or humidity control, it can lead to other problems such as condensation, insects, dust, pollen, mold, and even frozen pipes during the winter months. Special attention must be used in areas of your facility that house your cooking processes. Many of these processes rely on combustion of fuels for their heating source. Each combustion process requires a specific quantity of air for complete combustion of the fuel to occur. If the area is starved of outside air you will waste fuel, potentially add unwanted products of combustion to your process and, at a certain point, create an unsafe environment with the potential for an explosion to occur. Starving these cooking process areas of the required outside air can also affect the ability of the process controls to maintain temperature and humidity as specified in your HACCP plan. The best plan is to build it tight and ventilate it right. Then you will have a controllable air environment in your facility.

Facilitate Sanitation

Principle 6: Site Elements Facilitate Sanitary Conditions

Your property line is your facility's control perimeter. It demarcates the point at which you can exercise control. The first thing to consider is the general appearance of your property. The public's perception of your company and your brands can be influenced by the appearance of your property. What would the public's reaction be to pictures of your property on the nightly news? Your employee's perception is further imprinted by the habitual routine of showing up for work on a daily basis. If they arrive to a trash-strewn parking lot, overgrown grass, and cigarette butts around the smoking areas it will affect their attitude toward their assigned tasks in your facility. Poorly maintained property is the direct result of poor discipline. Poor discipline is fatally detrimental to your food safety program. Your HACCP program requires disciplined repeatability to control the surprises.

In addition to instilling the right impression, good site design should provide control of water flows, pests, dust, and dirt. I like to begin the site design process by asking how I can attract potential pests away from my buildings. In general, it costs a lot less to deal with potential contamination sources outside of your building and the cost continues to escalate as you move into your higher hygiene areas. It just makes sense to get the site design right.

Standing water on your site is an invitation to pests and other wildlife that can become vectors that spread potentially harmful microorganisms to your facility or employees. Your design must have properly sloped surfaces to direct water flow to a stormwater system. If you are required by regulation to have a stormwater detention pond on your site, locate it as far as practical from your buildings.

Birds, rodents, and insects are the major pests that your design has to control. Your site design must deprive these pests of any harborage. Tall grass or other uncut vegetation, trees that provide roosting sites or provide food in the form of fruit are invitations for unwanted guests. All grass areas need to be designed to be closely mown. This requires your design to limit the slope of any hill or embankment to a maximum of a 3:1 slope for safe mowing. Tree species should be selected that have the least potential to provide harborage and they should never be planted in close proximity to the facility. Outdoor lighting sources should be installed away from the building and be aimed toward the building. This will attract night-flying insects toward the light source and away from your building. A 3-foot wide rat run should be provided along the building's foundation walls to deprive rodents of cover, provide a space that can be easily monitored for rodent activity, and provide a hard stable surface for rodent bait stations.

Dust and dirt on your site are controlled mainly through proper paving and the installation of lawn areas. These areas then need to be properly maintained to continue to provide the design benefit. Failure to control dust and dirt will result in it being transferred into your facility by employees or infiltration through open doors and cracks or openings in your building envelope. It will also cause premature loading of your make-up air unit filters.

Principle 7: The Building Envelope Facilitates Sanitary Conditions

The building envelope consists of your foundation, walls, roof, and all of the openings that we install in them. The building envelope functions much like your skin. It is designed to keep in what you need to keep in and keep out what you do not want in. Your building's envelope has a very large surface area—hundred of thousands of square feet. Yet, your goal is to protect your facility from dust, water, insects, and rodents that are very small. A good designer needs to possess a built-in mental zoom function and focus on the detail level in order to properly design your building envelope. A mouse only needs an opening the size of a dime to gain access to your facility. Your focus should be on all building envelope penetrations such as air in-takes, exhaust outlets, doors, dock levelers, windows, pipes, and ducts. Are they properly flashed and sealed? Are bird and insect screens in place? A professional from your pest management company is a valuable second set of eyes for auditing your building envelope for defects that may allow pests into your facility.

Moisture control is another primary function of your building envelope. The building envelope must control water in its liquid and vapor form. The need for a good roofing system is evident to everyone. Unfortunately, the concept of water vapor flow is not so evident. It is extremely important that your design and construction team is experienced with constructing refrigerated buildings and is thoroughly knowledgeable with the design of vapor barriers for these applications.

Your building envelope is not a one-time investment. It will need to outlast many line changes and reconfigurations over its useful life. Roofing, sealants, door seals, and screens all need to be audited and maintained. Obsolete equipment and pipes should be removed and the building envelope properly repaired.

Principle 8: Interior Spatial Design Promotes Sanitation

This principle is fairly straightforward. Does your building design and equipment layout provide adequate space to safely access equipment and building components for sanitation and maintenance? Your design needs to provide space for full door swings and panel removal for all equipment. It also needs to factor in the type of equipment that may be necessary to safely maintain and clean the equipment and structure. Large horsepower motors and large parts require hoists or fork lifts to safely remove or install them. Overhead structures, light fixtures, refrigeration units, or air distribution plenums need to be easily accessible with ladders, scaffolds or man lifts. This requires adequate floor space to set up and properly position these devices for safe access. The temptation is to squeeze one more line into an existing space. In such circumstances the impact on sanitation and maintenance needs to be thoroughly reevaluated to make sure you are not creating future problems.

Principle 9: Building Components and Construction Facilitate Sanitary Conditions

This principle is implemented through the specifications and detail sections of your construction documents. The goal is to eliminate niches, harborages, and other food safety hazards through proper specification of materials, finishes, and sanitary design details.

All materials and finishes specifications must be able to withstand the normal wear and tear that is to be expected in a food processing facility without deterioration that could create a food safety hazard. This will require you to examine the potential impact of sanitation chemicals and processing agents such as brine as well as the mechanical forces from high-pressure hoses, temperature differentials, fork lifts, pallets, and vats on your material selections. Deteriorating floors and walls are well-documented harborage spots for pathogens. Plant shutdowns to remedy these situations are very costly in terms of lost production and construction costs and also create new food safety risks to ongoing operations during the construction repair process so it is wise to make sure that your specifications are fairly robust.

Proper design details improve the cleanability of your facility. You should avoid sharp angles in favor of coved surfaces at critical spots like your floor and wall juncture. This will make it easier to clean these areas. You should also avoid flat surfaces as they hold dirt and moisture since they are inherently not free-draining thereby creating perfect growth niches for microorganisms. Caulk joints need to be properly designed in order to function properly under the normal expansion and contraction forces they are subject to. Failed caulk joints typically hold moisture and are perfect niches for microorganisms.

Principle 10: Utility Systems Are Designed to Prevent Contamination

Utility systems are necessary for any facility to function. However, from a food safety perspective they can pose quite a number of problems. Porous insulation, materials that are not noncorrosive, hangers, brackets, and fittings all can prove to be food safety hazards if not designed to prevent niches and harborages and foreign material contamination. First and foremost, these systems need to be designed to withstand the sanitation process. Second, they need to be designed such that they are never overexposed product. Third, they need to be designed to allow adequate space between the pipes and between building surfaces to allow for inspection and sanitation. The best solution is to isolate as much of the utility system infrastructure from your processing environment as you can. A walkable ceiling structure can effectively eliminate any horizontal infrastructure runs over your product areas and should be considered in your facility design. Figure 10.1 shows a very effectively executed clean packaging room that complies with this principle.

Electrical systems pose particular issues in cold, wet environments. Improperly sealed electrical conduits and fittings can create moisture pathways into your electrical control cabinets which are nice, cozy warm places in your otherwise cold environment. For microorganisms these spots become a favorite destination—sort of like our trips to the islands. Unfortunately, every time your maintenance technician reaches into the control cabinet to troubleshoot a problem his hands and tools come out contaminated. Needless to say, this situation also creates a personnel safety issue.

Principle 11: Sanitation Is Integrated into Facility Design

This is common sense. You would not buy a stripped down car and take it to one vendor to add air conditioning and another vendor to add the radio, etc. Can it be done? Sure, but the cost and results are unlikely to be the same as buying the package completely integrated from one vendor. Then why does it still happen when it comes to sanitation? I think there are two primary reasons. First, few architectural/engineering firms have

Figure 10.1. A well-designed clean room.

sanitation expertise. Second, few companies actively involve their sanitation managers and sanitation chemical suppliers in the design process. It is normally turned over to them after the construction is completed. Integrating the sanitation process into the facility and equipment design requires intimate knowledge of the processes. Likewise, the sanitation processes can have significant impacts on other design decisions. Material selection must factor in the specified sanitation chemicals. Wastewater treatment and environmental compliance must also factor in the impact of sanitation chemicals. So, it just makes sense to integrate the sanitation process into your design efforts from day one.

There are four critical design aspects to achieving the integration—infrastructure capacity, spatial layout, safe access, and time. Your infrastructure capacity needs to be capable of providing the correct temperature and volume of water required at each individual device. When you are working in a refrigerated environment, cold water is a large disincentive to following prescribed hand washing procedures. Properly sized piping systems will also save you a lot of energy by reducing unnecessary pressure drop throughout the system.

Good spatial layout is critical to both your food safety and personnel safety programs. Some sanitation equipment, such as vat washers are quite large and require dedicated space to accommodate their footprint but they also require adequate forklift maneuvering area and storage for both clean and dirt vats to prevent cross-contamination. This is difficult to achieve as an afterthought in the design process. Hose stations also need to be correctly laid out to provide ease of access to every building element and piece of equipment. Hose station placement should never require a hose to be draped over a piece of equipment in order to reach another piece of equipment. After all, hoses lay on the floor and the floor should always be considered as a contaminated surface so we do not want hoses to touch equipment since this can be a transfer vector for contaminants.

Your design also needs to factor in the need for safe access to clean all surfaces. Ladders, scaffolds, and power lifts will likely be required to safely reach all surfaces for either daily or prescribed periodic cleaning. Your design needs to provide adequate access, setup, and maneuvering space to safely achieve this. If it does not, it is highly likely that the job will not get done correctly or safely.

Finally, you and your design team need to agree on a targeted time frame for the sanitation process. This decision impacts the economics of your facility as well as the design approach your team will need to take. As the allowed sanitation time decreases, the production capacity of your facility increases. This is a good outcome which can dramatically impact your facility's return on invested capital. However, it will also dictate some design changes to allow a faster sanitation cycle. It may require more sanitation manpower which will mean more hoses being used at one time which therefore requires more infrastructure capacity to handle the increased flows. It may require further enhancements to your refrigeration system to remove more water vapor and dry the rooms out in a shorter period of time. It may also require you to invest more in specific cleaning systems such as CIP and COP systems. These options can all be analyzed against your company's financial guidelines to determine whether they are good investments for your particular situation.

It is very difficult to maintain sanitary design concepts over the life of a facility unless you put an audit process in place to identify issues as they arise and create action plans to correct the issues. Normal expansion and contraction due to annual temperature variations along with normal wear and tear of the surfaces of your facility can create enough harborage points to threaten your controllable environment. Once this occurs, you are at a much greater risk of a product contamination problem. Every activity in your facility tends to lead to your production equipment. Your employees travel back and forth to their assigned stations on your production lines, maintenance personnel travel back and forth to your production lines, quality assurance personnel go to your production lines to collect their samples. The list could go on but the point is that a harborage location in your facility is highly likely to become a transfer point to your employees who will then carry the microorganisms to your production lines simply because your production lines are their destinations.

CHAPTER 11

Third-Party Audits

Robert E. Rust, Iowa State University

What Is a Third-Party Audit?

It is a means to provide an independent determination that a food-processing operation has appropriately designed systems and is operating within these systems so as to insure the safety, quality, security, and consistency of the food product(s) being produced. While compliance with appropriate regulatory standards will be part of the audit, the audit differs from inspection in that it should cover all aspects of the operation that contribute to, not only safety, but quality and consistency as well.

A third-party audit may be required by a customer or by management, essentially an internal audit. An internal audit is extremely useful in insuring consistency between plants in a multi-plant organization.

The audit essentially provides an opportunity for a completely independent view of the food-processing operation. It will document existing programs and conditions as well as compliance to regulations and manufacturing standards. It improves food safety and quality awareness in the plant and goes a long way in assuring customer trust. The results of the audit provide a benchmark for future improvement. It should also be looked on as a learning tool for all of the personnel and definitely as a vehicle for improvement.

A third-party audit should be unbiased and conducted by someone who is independent of the organization. The auditor should be well trained in the specifics of conducting an audit and may, in fact, be certified by a recognized certifying body. In addition, this person should be thoroughly familiar with the technology involved in the particular product production as well as the regulatory standards that apply to the food product produced.

Preparing for an Audit

Third-party audits are usually scheduled. Depending on complexity you should be prepared for them to last one to two days. These details can be worked out in advance of the audit. Many auditing firms will provide a manual, training programs, or other forms of guidance to assist in maximizing the productivity of an audit.

Since the auditor will be working from a prescribed document that assigns specific points to be audited, that document should be available well in advance of the audit. This provides time to assemble the required documentation. There is nothing more frustrating and unproductive than having the auditor muddle through a heap of poorly organized documents. All of the key people that the auditor needs to interview should be readily available. It is also important that the plant be in full production at the time of the audit so that the auditor can verify that the specified food safety and quality standards are actually being applied in production.

A person or persons should be assigned to work with the auditor to help locate and interpret documents should questions arise. The assigned individuals should keep notes during the audit to serve as an aid in improving the effectiveness of future audits.

Areas Covered in an Audit

Administration and Regulatory Compliance

1. Food safety, quality and security organization, and responsibilities of personnel involved. There should be well-documented chains of command and responsibilities of persons involved should be clearly delineated.
2. Food safety, quality and security policies, and procedures. These should be appropriately documented and the documents readily available to all personnel.
3. Specific training programs for management and operating personnel. Certainly, there should be well-defined goals and expectations for these training programs. There should be well-documented training programs for new hire personnel. Involvement in on-site training should be documented and examples of training materials should be available. There should also be records of personnel participation continuing education programs.

4. Recall plans and procedures. In addition to the appropriate documentation, results of mock recalls would be included in any review of this area.
5. Regulatory compliance. This can be determined from a review of all inspection documents as well as interviews with personnel directly involved in the inspection process.
6. Documents and records management. This would involve not only organization of the documents but also the archiving and eventual disposal of past records.
7. Change management. Who is authorized to "sign off" on changes in formulations or production practices, for example? How are changes initiated? Are changes well documented?
8. The awareness of management and commitment to food safety and quality. This should include the tracking of the effectiveness of company policies. An interview with the key management personnel will help the auditor get a clearer picture of this aspect.
9. Crisis and natural disaster management. Not only should a well-documented set of procedures be in place but also these should be routinely tested to insure functionality.
10. Customer and consumer complaints. Again, is there a clearly delineated response to the complaint issue? An auditor should also review not only the response but also the volume and frequency of complaints. Do complaints reflect a lack of control within the system?

HACCP Management

The entire HACCP program is obviously one of the key issues in any food safety audit. It should start with the basic issue of how the HACCP program was constituted. Minutes of HACCP team meetings should be reviewed, as should any HACCP training for individuals within the operation. What steps are being taken to constantly review and update the program? Expect the auditor to provide a critical review even though the HACCP plans and implementation will already be under constant review by the appropriate inspection authority.

Facility and Equipment Review

This will include both a "paper" review as well as a physical observation of the premises. Here are some of the areas of emphasis:

1. Overall plant design and construction. One thing not to be overlooked is the updating of design blueprints as modifications are made. Water systems, both potable and wastewater, should receive a critical review. With today's emphasis on conservation, adequate attention to this detail is one sign of a well-managed and controlled operation. A review here should also include a critical review of equipment layout and product flow.
2. The existing condition of the plant and equipment. How well is the plant maintained and is scheduled preventative maintenance being accomplished. Are adequate maintenance standards in place and do these support the key programs such as sanitation, general housekeeping, etc.?
3. The ready-to-eat production areas need special attention. Are these areas designed and operated in such a manner as to minimize or even eliminate chances for postprocessing contamination?
4. Employee support facilities. Not only employee welfare facilities should be reviewed but the in-plant availability and functionality of such facilities as hand washing, footwear sanitation, and uniform changing need attention. A long-time management person in the meat industry once stated that he could tell a lot about the operation simply by observing the condition of the locker rooms and the plant maintenance shop. If these were in good condition then you could almost wager that the rest of the operation was under good control.
5. Plant lighting and protection. Is lighting adequate, particularly in those areas where critical operations are being performed?

Sanitation, Housekeeping, and Hygiene

Certainly a review of standard sanitation operating procedures (SSOPs) is in order. Not only should the SSOPs be in place but the monitoring of these should also be well documented. Since cleaning and sanitizing chemicals are a potential hazard, the appropriate control of these is important. This should include the training of sanitation personnel in appropriate material use. Is there verification of the effectiveness of cleaning and sanitizing procedures and is this well documented?

Since, in meat processing operations, preoperational inspection is a required procedure, this should be well documented along with any corrective actions. A high frequency of corrective actions during preoperation inspection could indicate that a review of sanitation or maintenance needs to be instituted.

Personal hygiene should be observed carefully. Adherence to specified dress codes for certain areas of the plant is one area of special attention. Along with this, policies regarding restricted entry into critical areas should be addressed.

Rodent and Pest Control

This needs to extend beyond the wall of the production facility. All control programs need documentation of both the controls being instituted as well as observed pest activities. Since this is an area often delegated to outside contractors, there should be insistence on detailed documentation of the services performed.

Because pest control chemicals are a potential hazard in themselves, appropriate storage and control need to be exercised. This includes training of personnel authorized to use these chemicals.

It goes without saying that facility design and construction needs to such that it will minimize rodent and pest entry.

Receiving and Inventory Control

An appropriate safety and quality control program must include controls at the receiving dock with specific policies, inspection, and acceptance plans. Incoming vehicles require review and documentation. There need to be specific release criteria for incoming materials. Storage and handling policies and practices need documentation. Attention to policies for the handling and sanitation control of bulk materials should be well documented. Sensitive ingredients including chemicals and restricted ingredients require appropriate control and documentation. To facilitate traceability, all incoming materials need to be identified, documented, and appropriately inventoried.

Process and Product Evaluation

To appropriately audit this phase, the auditor needs to combine an examination of control documents and a careful observation of the actual manufacturing process. This is why the plant needs to be in full operation when the audit is scheduled. If multiple shifts are involved, the auditor should observe production at each shift. Specific points to cover are:

1. Process control procedures and documentation.
2. Product specifications, formulation control, and the accuracy of the formulation. This should include documentation that insures that the product is consistently meeting specifications.
3. Routine calibration and preventative maintenance for all operation and measurement devices involved in production and product control.
4. Control of foreign materials and potential contaminants.
5. Statistical controls and the appropriate documentation of these controls.
6. Controls over sensitive and restricted ingredients, particularly potential allergens.
7. Handling and documentation of rework and carry-over products. Particular attention should be paid to how and when the rework cycle is broken and how well this is documented.
8. The management and application of analytical data.

Packaging and Labeling

The audit should include all of the aspects of proper labeling and label control.

1. Label accuracy and compliance with regulations and customer specifications.
2. Documented net weight or product count compliance. Performance can be monitored by observation of appropriate statistical control documents.
3. Clarity and accuracy of manufacturing codes on individual and cased products.
4. Functionality and integrity of packaging. This should include documentation of package performance tests.
5. Label security and control of obsolete labels.
6. Use of tamper-evident packaging where appropriate.

Storage and Shipping

In this case the audit needs to be extended to commercial warehousing facilities if they are used. Points to be covered are:

1. Warehouse and finished product management.
2. Control and handling of retained and returned goods.
3. Maintenance and management of storage areas and shipping docks.

4. Condition, control, and management of transport equipment. This should include monitoring of controls exercised during transport.
5. Authorization to release and ship product.
6. Traceability throughout the system.

Laboratory Support

A particular operation may use an on-site laboratory or may utilize the services of a commercial laboratory. In either case the audit should include a careful assessment of the adequacy of the laboratory and the qualifications of the staffing. Procedures should be standardized and recognized as appropriate. These also need documentation. Equipment calibration needs to be verified and accuracy of the laboratory analyses needs periodic verification by recognized methods.

Product Security

The issue of product security needs to be addressed. How is security managed? Are the facilities and the operations secure? What is being done to control the human element as it relates to security? Are industry or regulatory guidelines being followed?

Finalizing the Audit

This is the time when the auditor meets with all of the involved personnel in the management team to present his/her findings. This should be viewed as a critical learning experience. If the auditor has found deficiencies or areas where improvement can be made, these should be thoroughly discussed. While the auditor may suggest some ways that improvement can be effected, do not expect the auditor to fill the role of a consultant. If in-house expertise is not available to make necessary improvement, it may be necessary to bring in an outside consultant to provide technical assistance. Again, do not look at a third-party audit as something to fear. Look at it as a learning tool for improvement. To paraphrase Robert Burns, the immortal Scots poet, "Would that we had the power to see ourselves as others see us."

CHAPTER 12

Food Safety Beyond Guidelines and Regulations

Bradley P. Marks, Michigan State University

Introduction

Growth in the ready-to-eat (RTE) market sector and evolving federal regulations are creating a need for better information related to inactivation and growth of pathogens in meat and poultry products. Regulatory changes are shifting the burden to processors to ensure, *through scientific rationale*, that a new or modified process meets lethality and stabilization performance standards. Although product and process parameters are known to affect thermal resistance of bacteria, most reported information is from laboratory studies that encompass a limited range of conditions. In most cases, the validity of this information for commercial processes is uncertain. Therefore, the purpose of this chapter is to address three questions: (1) How does the scientific domain of the performance standards for RTE products relate to the current state of knowledge? (2) What is currently known about various factors that might affect thermal inactivation of pathogens in meat and poultry products? (3) What should be done to account for these complicating factors, now and in the future?

Relating Regulations and Guidelines to the State of the Art

Thermal processing is the primary method for both adding value and ensuring microbial safety of meat and poultry products. Although numerous technologies (e.g., irradiation, ultra-high pressure, and pulsed electric fields) loom on the horizon for the broader food industry, the application of heat will certainly continue as the dominant means to impart desirable characteristics, add economic value, and ensure product safety. Additionally, major shifts in consumer demand and regulatory burden are increasing the importance of thermal processing in all of these areas. Therefore, this chapter focuses on thermal processing as the key step in ensuring the safety of RTE products.

Regulatory Evolution

In terms of regulatory pressures in this domain, there is an evolving shift from a command-and-control paradigm (i.e., meeting specific endpoint temperatures) to lethality performance standards. This shift is evident in one final rule (FSIS, 1999) and one proposed rule (FSIS, 2001a). The 1999 rule changes specified lethality performance standards for cooked and roast beef (CFR §318.17) and cooked poultry products (CFR §381.150), but continued to provide only time–temperature specifications for cooked patties (CFR §318.23). The amended regulations state that any process producing RTE, whole-muscle products must achieve 6.5 $\log_{10}$ or 7.0 $\log_{10}$ reduction in *Salmonella* for whole-muscle beef or poultry, respectively.

A central theme to these regulatory changes is the emphasis on developing science-based regulations. For example, processors are no longer held to specific endpoint temperatures; however, they "must validate new or altered process schedules by scientifically supportable means" (FSIS, 1999). Appendix A, the compliance guidelines for meeting the lethality performance standards (FSIS, 1999), states that processors may develop customized procedures that meet the performance standards by ". . . using information obtained from the literature." Alternatively, the same guidelines suggest that an inoculated challenge study (adding pathogens to a real food product prior to processing) is a "definitive tool" for ensuring process lethality, and should be carried out with ". . . a cocktail of various serotypes of *Salmonella* that have been historically implicated in outbreaks."

The proposed rule changes (FSIS, 2001a) extend the same general approach to all RTE products containing meat or poultry. The proposed changes state that a processor will need to ". . . demonstrate the relationships between the lethality treatments and the specific characteristics of a product, such as physical and chemical properties." The demonstration ". . . could involve the use of heat transfer equations and should account for all variables that would affect lethality (e.g., size of product, humidity, density, thermal conductivity, specific heat, shape, product composition, and strain of organism)." Lastly, the proposed changes also state that even if a process meets the above lethality performance standards, "... any detectable levels of viable *Salmonella* ... would render that product adulterated."

However, a recent FSIS notice (FSIS, 2002) to inspection personnel stated that, "... it is not possible or appropriate to rely solely on a predictive modeling program to determine the safety of foods ...." In that notice, FSIS spells out an expectation for either experimental validation of any predictive model or a documented decision from an expert (microbiologist) that the model predictions are sufficiently safe. Additionally, the notice comments on the fact that very few microbial models account for all the relevant process and product parameters affecting microbial behavior (which is discussed later in this chapter).

The 1999 rule changes mentioned above do allow for processing within "safe harbor" guidelines (i.e., specified time–temperature combinations), and draft compliance guidelines have been published for certain common products covered by the currently proposed changes (FSIS, 2001b). However, there are two major and foreseeable problems. First, it is unlikely that compliance guidelines will be published for every affected product, particularly niche specialty and ethnic products, thereby leaving many processors (probably those with the least likelihood of having in-house capacity to address the issue) without "safe harbors" and forcing them to prove that their processes meet the lethality performance standards. Second, the current evolutionary shift of the microbial safety burden for RTE products to processors may likely foreshadow a more complete change, given that there are no guarantees that the "safe harbor" guidelines will remain in place indefinitely. Although the new regulatory paradigm creates greater opportunities for customized processes, it clearly puts significant pressure on the industry to document process lethality for any new product or process.

State of the Art

If the state of the art encompasses two domains, *knowledge and tools*, then it is important to assess the intersection of these domains with the evolving regulatory domain described above (Fig. 12.1). In this case, the *knowledge* base consists of the latest research related to pathogen response to processing, product effects, etc., which will be discussed in the next section. As will be seen, a fair amount of previous research has

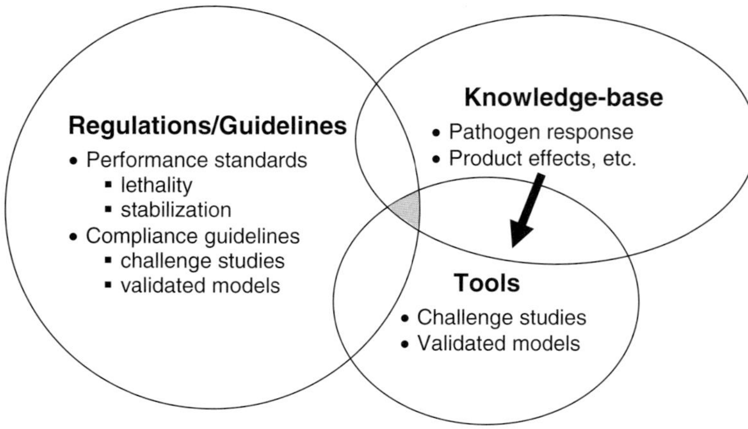

Figure 12.1. The intersecting domains of food safety regulations, scientific knowledge, and validated tools based on that knowledge.

been aimed at developing this type of knowledge. However, mere knowledge that these effects exist is insufficient to aid a processor in designing, operating, or evaluating the efficacy of a thermal process, in terms of the relevant lethality performance standards. Quantitative analysis requires *tools*. The relevant tools are either direct measurement via inoculated challenge studies (relevant only for validating existing processes) or validated models that can predict the lethality outcome for specified product and process conditions (relevant for either designing new processes or evaluating existing ones). Operating in the area where the regulatory and knowledge domains overlap may result in food safety decisions that are essentially "educated guesses." However, if that knowledge has been used to develop validated tools, then operating in the area where the regulatory and knowledge-based tool domains overlap should result in reliable assessments of process lethality and product safety.

Unfortunately, the current state of the art, regarding validated tools, is insufficient for *100% reliable* lethality predictions in commercial processes. The strengths and weaknesses of the two general methods (challenge studies and predictive models) are listed in Table 12.1.

In general, challenge studies (i.e., inoculation of real products with target organisms) are impossible in commercial facilities, where pathogens cannot be brought on site. If the capacity does exist to conduct such tests, they have the advantage of directly accounting for any product effects by virtue of using the actual product in the tests. However, it is important to understand that direct matching of the conditions in a pilot-scale test with those in an actual commercial process can be extremely difficult, due to process scale-up issues.

Table 12.1. Strengths and weaknesses of tools for validating thermal process lethality

	Challenge Studies	*Predictive Models*
Strengths	1. Product-specific results 2. Results are lumped for an entire piece of product (i.e., the real case)	1. No requirement of special biohazard facilities or testing 2. Can compare effects of using different literature data
Weaknesses	1. Not practical (due to typical lack of biohazard pilot-processing facilities) 2. Results are strain-dependent 3. It is difficult for off-line tests to exactly mimic actual process 4. Pathogen recovery methods affect results (e.g., sublethally injured cells)	1. Typically based on only T_{center}, which neglects the "lethality profile" in a product 2. Few account for factors other than time and temperature 3. Usefulness limited to domain used for model validation 4. Unlikely to find inactivation models exactly matching specific product/process scenario

Additionally, existing pathogen inactivation data and models (as described in the next two sections) have been developed primarily with pathogen cultures grown under ideal laboratory conditions, inoculated into model products, and subjected to isothermal laboratory conditions; therefore, the resulting models are not necessarily valid for conditions occurring in many commercial processes. In fact, one of the greatest dangers in using predictive models from the literature is extrapolation to conditions for which the model has not been validated. When this is done, the reliability of the prediction is impossible to quantify. Therefore, "scientifically supportable means" currently do not exist for *100% reliable and robust* predictions of thermal process lethality in this industry.

Factors Affecting Thermal Inactivation

Obviously, heat inactivates bacteria. However, when evaluating thermal process lethality, it is essential to understand the wide range of factors, beyond just temperature, that affect the process outcome. These factors can be classified as pathogen, product, or process parameters. This section briefly summarizes the state of knowledge in this area.

Pathogen Factors

Thermal resistance of pathogens can vary widely, depending on the organism being considered. *Salmonella* was selected as the reference organism for the lethality performance standards in part because it tends to be more thermally stable than other bacterial pathogens of concern (FSIS, 2001a). Even among various strains of *Salmonella*, there can be a significant difference in thermal resistance (Murphy et al., 1999).

Once the pathogens are in a food matrix, the previous conditions to which they have been subjected can also significantly affect their future response. For example, microbial stress is a physical or a chemical condition not severe enough to kill, but resulting in sublethally injured microbes (Hurst, 1977; Murano and Pierson, 1993). Potentially stressful factors include acid shock, sanitizer stress, oxidants, salt, starvation, heat shock, ionizing radiation, freezing, and thawing (Abee and Wouters, 1999; Lou and Yousef, 1996; Miller et al., 2000). As a result of sublethal stress, bacteria may exhibit enhanced thermal resistance (Bunning et al., 1990; Farber and Brown, 1990; Lou and Yousef, 1996; Mackey and Derrick, 1986; Pagán et al., 1997; Taormina and Beuchat, 2001; Wesche et al., 2005). As a result, a pathogen cell on a carcass, which is held chilled before further processing, might thereby develop increased thermal resistance due to the cold stress. Likewise, pathogen cells that are exposed to, but not inactivated by, sanitizing agents, could thereafter also exhibit greater thermal resistance. In general, a wide range of environmental stresses might result in indigenous populations of pathogens having greater heat resistance than observed in laboratory studies. Although stressed cells respond differently than cells grown under optimal conditions, it is the latter cells that are routinely used in food safety studies. Therefore, process validations based on such studies may not be adequate to ensure the safety of RTE products.

Product Factors

In terms of product attributes, the heat resistance of pathogens can be affected by meat species, muscle type, pH, carbohydrates, fat content, and salts (Jay, 1996). For example, thermal resistance of pathogens tends to increase with increasing fat content in the substrate (Ahmed et al., 1995; Juneja et al., 1997; Line and Harrison, 1992; Maurer, 2001; Veeramuthu et al., 1998), although not all researchers have seen this effect (Fain et al., 1991; Kotrola and Conner, 1997; Young et al., 1991). In any medium, there is typically an optimal pH for maximum bacterial heat resistance (Chiruta et al., 1997; Gaillard et al., 1998; Juneja et al., 1995b; Juneja and Eblen, 1999; Reichart, 1994). Additionally, additives such as salts, lactates, and phosphates may enhance thermal resistance of pathogens (Kotrola and Conner, 1997; Maurer, 2001).

Regarding the largest chemical component, water activity (and/or moisture content) of the product is widely known to be a controlling factor in microbial growth; however, its effect on thermal inactivation is less obvious, but extremely important. Blankenship (1978) and Goodfellow and Brown (1978) reported *Salmonella* survivors on the surfaces of fully-cooked, dry roasted beef and suggested that thermal resistance was enhanced by the reduction in water activity (a_w) near the meat surface. However, very few studies have quantified the effects of moisture content or a_w on microbial inactivation. Kirby and Davies (1990) dehydrated cultures of *Salmonella* Typhimurium and reported an increased thermal resistance; however, they were heating pure cultures rather than a food product. Corry (1975) had previously reported, but not modeled, the dependence of decimal reduction time (*D*) on a_w for *S.* Typhimurium and *S.* Senftenberg. The authors reported that thermal inactivation was highest at high a_w (>0.95), decreased with decreasing a_w until it reached a minimum between 0.6 and 0.8 a_w, and subsequently increased again as a_w approached zero. In evaluating this effect in ground turkey meat, Carlson et al. (2005) reported that the rate of thermal inactivation of *Salmonella* decreased 64% when a_w was decreased from 0.99 to 0.95.

In general, thermal resistance of bacteria is higher in meat products than in buffer solutions, peptone, agar, or other model systems (Bell and DeLacy, 1984; Blankenship and Craven, 1982; Ghazala et al., 1995; Juneja et al., 1995a, 2001; Murphy et al., 1999). Although a few researchers have reported *D* values for various pathogens in specific meat products (Blankenship and Craven, 1982; Fain et al., 1991; Murphy et al., 1999, 2002; Veeramuthu et al., 1998), very few have quantitatively modeled the relationships between product parameters and pathogen inactivation rates.

Not only do food components appear to enhance heat resistance, but cell location (surface attachment vs. interior dispersion) may also affect the resistance of *Salmonella* (Doyle and Mazzotta, 2000). Specifically, the thermal resistance of *Salmonella* in whole-muscle beef, pork, and turkey is significantly greater than in ground product of equivalent composition (Orta-Ramirez et al., 2005; Tuntivanich et al., 2005; Velasquez et al., 2005). Therefore, the validity of applying previous inactivation data from liquid media or meat slurries to thermal process calculations for real meat and poultry products is not well known.

Process Factors

Although product factors, such as pH, fat content, and water activity, all affect the thermal resistance of pathogens, the environmental conditions during thermal processing can affect the inactivation of pathogen both directly and indirectly, by changing the product properties during processing and controlling the rate of temperature change. In this context, consider process factors to be those parameters that can be controlled either by the process design or operation, such as heating (e.g., air) temperature, cooking time, humidity, and heating (or cooling) rates. With the exception of heating temperature and time, far less is known about the effects of environmental conditions on thermal inactivation, as relevant to commercial processes. As previously mentioned, these effects can be either immediate or delayed, as occurs when bacteria exhibit stress-induced tolerances to heat.

For example, although the effect of water activity on thermal resistance has been described, water activity is an intrinsic property of the product, reflecting the equilibrium vapor pressure over the product at a specified temperature. In a commercial impingement oven, high air velocities create a very small boundary layer around the product, so that the microenvironmental conditions around the product in the oven are essentially controlled by the oven conditions (i.e., temperature and humidity). Therefore, it may be the case that any inactivation mechanism for a pathogen cell at or near the surface of the product is thereby controlled more by the oven humidity than by the water activity of the supporting medium (i.e., the meat product). Murphy et al. (2001a) showed, via tests with chicken breast patties in a pilot-scale impingement oven, that air humidity had a significant effect on lethality, increasing inactivation of *Salmonella* by $>2 \log_{10}$ when comparing a high-humidity (steam-injected) treatment to a low-humidity (dry air) treatment to the same endpoint product temperatures. Murphy et al. (2001b) further showed that inactivation models based on isothermal laboratory inactivation studies in a water bath overpredicted *Salmonella* and *Listeria innocua* inactivation by as much as $5 \log_{10}$, when compared to experiments that heated chicken breast patties in a laboratory-scale (dry) convection oven. In Murphy et al. (2001a, 2001b), significant numbers of survivors were detected even after cooking to endpoint temperatures as high as 80°C in a dry convection environment. However, none of these previous studies isolated or modeled the effects of meat moisture content versus process humidity.

Additionally, there is sufficient evidence to suggest that thermal inactivation of pathogens in food systems is a path-dependent process. In other words, past handling and treatment affects thermal resistance of pathogens in future processes. For example, Stephens et al. (1994) tested the effects of heating rate on thermal inactivation of *L. monocytogenes* in tryptic phosphate broth. Contrary to the basic assumption behind kinetic-based modeling, bacterial inactivation decreased significantly for heating rates below 5°C/min, with deviations as high as 10^5-fold when comparing expected to observed survivor ratios. Subsequently, Stasiewicz et al. (2008) modeled this phenomenon, incorporating prior sublethal temperature history into a path-dependent model for the rate of thermal inactivation of *Salmonella* in ground turkey. Obviously, the importance of the heating rate effect depends on the type of thermal process and is more relevant to slow cooking procedures, where extending cooking time from 10 to 120 minutes can cause model prediction errors as large as 4 $\log_{10}$ (Mogollon et al., 2007).

What to Do (Now and in the Future)

Clearly, pathogen, product, and process parameters can have significant effects on the thermal resistance of pathogens in meat and poultry products. Because commercial cooking systems create complex conditions around the product, with varying temperature, humidity, airflow, etc., scale-up of laboratory-based inactivation data to commercial-scale processes, without evidence that the data account for all of the relevant process parameters, can be a risky leap. However, given the impracticality of challenge studies, the processor is left with predictive models as the primary means for evaluating and documenting process lethality. It is critically important, therefore, to determine the implications the aforementioned difficulties have for the present and for the future, in terms of process design, validation, and operation.

For Now

For the present, caution is the key recommendation regarding selection and use of published inactivation data and models. The most important caution about predictive models is that they should be validated, against data independent of those used to create the model, before they can be used for prediction of future results. Unfortunately, the vast majority of published data and models for thermal inactivation of pathogens are never validated as part of the original studies. Published models (including *D* values) are typically evaluated only in terms of their goodness of fit to the data used to

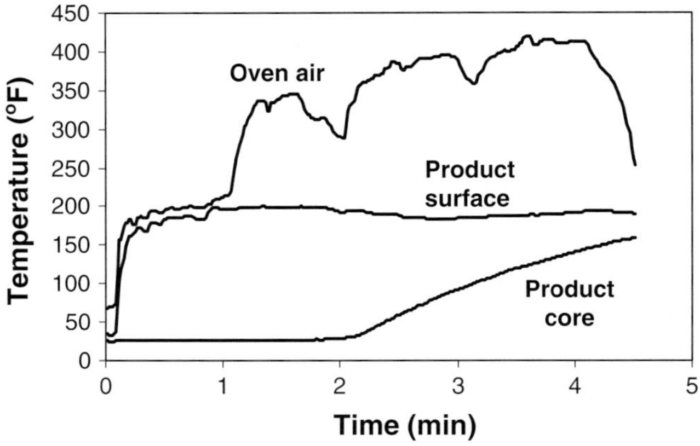

Figure 12.2. Example of thermal history used in process lethality calculations.

create the model. This type of analysis provides no information about the ability of a model to accurately predict the outcome of a future process.

As an example of the potential variability that can occur in process lethality calculations, consider the following illustration. Three different methods were used to predict the $\log_{10}$ reductions for *Salmonella* in a given cooking process, representing the thermal history for a ground and formed poultry product subjected to a multistage impingement oven system (Fig. 12.2). For the first method, a first-order inactivation model was applied, using the published kinetic parameters from Murphy et al. (1999), which were developed from a cocktail of six *Salmonella* serovars (Senftenberg, Typhimurium, Heidelberg, Mission, Montevideo, and California) heated in laboratory tests (60–70°C). For the second method, the AMI process lethality spreadsheet (AMI, 2001) was used, without modification, applying the "default" thermal resistance parameters previously listed for *Salmonella* in the spreadsheet. For the last method, a Weibull-based primary model (Peleg and Cole, 1998) with a square-root secondary model for the rate constant as a function of temperature was fit to raw inactivation data from a different *Salmonella* cocktail (Thompson FSIS 120, Enteriditis H3527 and H3502, Typhimurium H3380, Hadar MF60404, Copenhagen 8457, Montevideo FSIS 051, and Heidelberg F5038BG1), developed from laboratory inactivation trials (55–63°C) in ground turkey at Michigan State University (unpublished data). The root mean squared error for that model was ~0.7 $\log_{10}$. The resulting predictions, applied to the identical temperature data shown in Fig. 12.2, were ~3.7, 29, and 96 $\log_{10}$ reductions for the first, second, and third methods, respectively. All of the models appear legitimate, in terms of methodology, but the results were clearly widely different. The low value resulting from the kinetic model of Murphy et al. (1999) can be attributed to the fact that *S.* Senftenberg, a relatively very resistant strain, was included in that model. The huge number for the third case can be attributed to the fact that the square-root model, although it fit the original data well, was only valid for the stated temperature range (55–63°C), which was insufficient and inappropriate for the process that was considered ($T_{\text{final}} = 74°\text{C}$), which required extrapolation of the model. Finally, the validity of the second case would not be known to the user, given lack of information about the underlying uncertainty of the "default" parameters that were used.

Given such wide variability resulting from the use of different data and models, a processor needs to be extremely cautious in using published inactivation data. In particular, users of any microbial inactivation model should be sure to use parameters that most closely match their own situation, in terms of product type, fat and water content, process conditions, etc. Given that no universally applicable modeling tool yet exists, the best that the user can do is to consider comparing results from several different models and/or parameters that are most relevant to the specific case. This can help define a range of possible lethality outcomes. Most importantly, the user should particularly avoid extrapolating a given model to conditions beyond which the model has been validated, because there is no way to know the accuracy of the resulting predictions in this case.

For the Future

Clearly, there is a need for validated lethality models that have broad applicability across a range of products and processes, and which account for all of the factors known to affect lethality. With respect to inactivation model tools or programs, the compliance guidelines for the new FSIS regulations (FSIS, 1999) specifically refer to the USDA-ARS Pathogen Modeling Program (PMP, http://pmp.arserrc.gov), thereby inferring that this tool could be used to relate cooking parameters to pathogen lethality. Another example is the AMI process lethality spreadsheet (American Meat Institute, http://www.amif.org). Both are excellent examples of microbial modeling tools (simple to use, user-friendly, Windows-based); however, both have a number of limitations relevant to the real thermal processes. First, the current version of PMP does not include primary models for thermal inactivation of *Salmonella*, nor does either model include secondary models that relate important product and process conditions (e.g., fat content and humidity) to inactivation. Both assume first-order inactivation kinetics, which ignore any lag or tailing phenomena that can be important, in terms of resistant subpopulations. Additionally, neither model accounts for temperature and moisture gradients that occur in real food products and therefore cause "lethality profiles" within a meat product. Consequently, there is still a need to further extend the methods of quantitative microbiology to create "universal" *Salmonella* inactivation models for quantifying the lethality of commercial cooking systems.

Summary

The general observation that current microbial inactivation models fail to account for all of the factors relevant to commercial thermal processes is certainly of no comfort to an industry that is increasingly being compelled to verify and prove that cooking systems are meeting lethality performance standards. There is a significant need for user-friendly, publicly available, validated models that would allow a user to enter product conditions (composition, species, and structure) and process data (time and temperature) and get back a prediction of pathogen inactivation, including an estimate of uncertainty.

In the meantime, processors should be cautious in applying simple D and z values to integrated time–temperature histories from process data. Minimally, they should be aware of the medium and heating conditions used to generate the inactivation parameters, and recognize whether their processes differ from those conditions in significant ways, such as product composition or process humidity. Even though undercooking in food manufacturing facilities currently is not causing widespread food safety problems, continued development of new products and processes (and the ongoing regulatory changes) necessitate a proactive stance in ensuring proper evaluation of thermal process lethality.

References

Abee, T., and Wouters, J.A. 1999. Microbial stress response in minimal processing. *International Journal of Food Microbiology* 50:65–91.

Ahmed, M.N., Conner, D.E., and Huffman, D.L. 1995. Heat-resistance of *Escherichia coli* O157:H7 in meat and poultry as affected by product composition. *Journal of Food Science* 60:606–610.

AMI (American Meat Institute). 2001. AMI process lethality determination spreadsheet. Online Publication. Accessed at: www.amif.org.

Bell, R.G., and DeLacy, K.M. 1984. Heat injury and recovery of *Streptococcus faecium* associated with the souring of club-packed lunchen meat. *Journal of Applied Microbiology* 57:229–236.

Blankenship, L.C. 1978. Survival of *Salmonella typhimurium* experimental contaminant during cooking of beef roasts. *Applied and Environmental Microbiology* 35:1160–1165.

Blankenship, L.C., and Craven, S.E. 1982. *Campylobacter jejuni* survival in chicken meat as a function of temperature. *Applied and Environmental Microbiology* 44:88–92.

Bunning, V.K., Crawford, R.G., Tierney, J.T., and Peeler, J.T. 1990. Thermotolerance of *Listeria monocytogenes* and *Salmonella typhimurium* after sublethal heat shock. *Applied and Environmental Microbiology* 56:3216–3219.

Carlson, T.R., Marks, B.P., Booren, A.M., Ryser, E.T., and Orta-Ramirez, A. 2005. Effect of water activity on thermal inactivation of *Salmonella* in ground turkey. *Journal of Food Science* 70:363–366.

Chiruta, J., Davey, K.R., and Thomas, C.J. 1997. Thermal inactivation kinetics of three vegetative bacteria as influenced by combined temperature and pH in a liquid medium. *Food and Bioproducts Processing* 75:174–180.

Corry, J.E.L. 1975. The effect of water activity on the heat resistance of bacteria. In: Duckworth, R.B. (ed.), *Water Relations of Foods*. Proceedings of an International Symposium held in Glasgow, September 1974, pp. 325–337.

Doyle, M.E., and Mazzotta, A.S. 2000. Review of studies on the thermal resistance of Salmonellae. *Journal of Food Protection* 63:779–795.

Fain, A.R., Jr., Line, J.E., Moran, A.B., Martin, L.M., Lechowich, R.V., Carosella, J.M., and Brown, W.L. 1991. Lethality of heat to *Listeria monocytogenes* Scott A: D-value and Z-value determinations in ground beef and turkey. *Journal of Food Protection* 54:756–761.

Farber, J.M., and Brown, B.E. 1990. Effect of prior heat shock on heat resistance of *Listeria monocytogenes* in meat. *Applied and Environmental Microbiology* 56:1584–1587.

FSIS. 1999. Performance standards for production of certain meat and poultry products. United States Department of Agriculture. Food Safety Inspection Service. 9 CFR Parts 301, 317, 318, 320, and 381. *Federal Register* 64(3):732–749.

FSIS. 2001a. Performance standards for the production of processed meat and poultry products; proposed rule. United States Department of Agriculture. Food Safety Inspection Service. 9 CFR Parts 301, 303, et al. *Federal Register* 66(39):12590–12636.

FSIS. 2001b. Draft compliance guidelines for ready-to-eat meat and poultry products. United States Department of Agriculture. Food Safety Inspection Service. http://www.fsis.usda.gov/OPPDE/rdad/FRPubs/RTEGuide.pdf.

FSIS. 2002. Use of Microbial Pathogen Computer Modeling in HACCP Plans. FSIS Notice 55-02. U.S. Department of Agriculture. Food Safety Inspection Service, December 2, 2002.

Gaillard, S., Leguerinel, I., and Mafart, P. 1998. Model for combined effects of temperature, pH, and water activity on thermal inactivation of Bacillus cereus spores. *Journal of Food Science* 63:887–889.

Ghazala, S., Coxworthy, D., and Alkanani, T. 1995. Thermal kinetics of *Streptococcus faecium* in nutrient broth/*sous vide* products under pasteurization conditions. *Journal of Food Processing and Preservation* 19:243–257.

Goodfellow, S.J., and Brown, W.L. 1978. Fate of Salmonella inoculated into beef for cooking. *Journal of Food Protection* 41:598–605.

Hurst, A. 1977. Bacterial injury: A review. *Canadian Journal of Microbiology* 23:935–943.

Jay, J.M. 1996. *Modern Food Microbiology*, 5th edn. Chapman & Hall, New York.

Juneja, V.K., and Eblen, B.S. 1999. Predictive thermal inactivation model for *Listeria monocytogenes* with temperature, pH, NaCl, and sodium pyrophosphate as controlling factors. *Journal of Food Protection* 62:986–993.

Juneja, V.K., Eblen, B.S., Marmer, B.S., Williams, A.C., Palumbo, S.A., and Miller, A.J. 1995a. Thermal resistance of nonproteolytic type B and type E *Clostridium botulinum* spores in phosphate buffer and turkey slurry. *Journal of Food Protection* 58:758–763.

Juneja, V.K., Eblen, B.S., and Ransom, G.M. 2001. Thermal inactivation of *Salmonella* spp. in chicken broth, beef, pork, turkey, and chicken: Determination of D and z values. *Journal of Food Science* 66:146–152.

Juneja, V.K., Marmer, B.S., Phillips, J.G., and Miller, A.J. 1995b. Influence of intrinsic properties of food on thermal inactivation of spores of nonproteolytic *Clostridium botulinum*: development of a predictive model. *Journal of Food Safety* 15:349–364.

Juneja, V.K., Snyder, O.P., Jr., Williams, A.C., and Marmer, B.S. 1997. Thermal destruction of *Escherichia coli* O157:H7 in hamburger. *Journal of Food Protection* 60(10):1163–1166.

Kirby, R.M., and Davies, R. 1990. Survival of dehydrated cells of *Salmonella typhimurium* LT2 at high temperatures. *Journal of Applied Bacteriology* 68:241–246.

Kotrola, J.S., and Conner, D.E. 1997. Heat inactivation of *Escherichia coli* O157:H7 in turkey meat as affected by sodium chloride, sodium lactate, polyphosphate, and fat content. *Journal of Food Protection* 60:898–902.

Line, J.E., and Harrison, M.A. 1992. *Listeria monocytogenes* inactivation in turkey rolls and battered chicken nuggets subjected to simulated commercial cooking. *Journal of Food Science* 57:787–793.

Lou, Y., and Yousef, A.E. 1996. Resistance of *Listeria monocytogenes* to heat after adaptation to environmental stresses. *Journal of Food Protection* 59:465–471.

Mackey, B.M., and Derrick, C.M. 1986. Elevation of the heat resistance of *Salmonella typhimurium* by sublethal heat shock. *Journal of Applied Bacteriology* 61:389–393.

Maurer, J.L. 2001. *Environmental Effects on the Thermal Resistance of Salmonella, Escherichia coli O157:H7, and Triose Phosphate Isomerase in Ground Turkey and Beef*. M.S. Thesis. Michigan State University, East Lansing, MI.

Miller, A.J., Bayles, D.O., and Eblen, B.S. 2000. Cold shock induction of thermal sensitivity in *Listeria monocytogenes*. *Applied and Environmental Microbiology* 66:4345–4350.

Mogollon, M.A., Marks, B.P., Jeong, S., Stasiewicz, M.J., and Booren, A.M. 2007. Effect of cooking profiles and sub-lethal history on *Salmonella* thermal inactivation in whole-muscle beef. IFT Abstract 098-09. *Presented at the Institute of Food Technologists Annual Meeting*, Chicago, IL, July 2007.

Murano, E.A., and Pierson, M.D. 1993. Effect of heat shock and incubation atmosphere on injury and recovery of *Escherichia coli* O157:H7. *Journal of Food Protection* 56:568–572.

Murphy, R.Y., Duncan, L.K., Johnson, E.R., Davis, M.D., and Smith, J.N. 2002. Thermal inactivation D- and z-values of *Salmonella* serotypes and *Listeria innocua* in chicken patties, chicken tenders, franks, beef patties, and blended beef and turkey patties. *Journal of Food Protection* 65:53–60.

Murphy, R.Y., Johnson, E.R., Marcy, J.A., and Johnson, M.G. 2001a. Survival and growth of *Salmonella* and *Listeria* in the chicken breast patties subjected to time and temperature abuse under varying conditions. *Journal of Food Protection* 64:23–29.

Murphy, R.Y., Johnson, E.R., Marks, B.P., Johnson, M.G., and Marcy, J.A. 2001b. Thermal inactivation of *Salmonella senftenberg* and *Listeria innocua* in ground chicken breast patties processed in an air convection oven. *Poultry Science* 80:515–521.

Murphy, R.Y., Marks, B.P., Johnson, E.R., and Johnson, M.G. 1999. Inactivation of *Salmonella* and *Listeria* in ground chicken breast meat during thermal processing. *Journal of Food Protection* 62:980–985.

Orta-Ramirez, A., Marks, B.P., Warsow, C.R., Booren, A.M., and Ryser, E.T. 2005. Enhanced thermal resistance of *Salmonella* in whole muscle vs. ground beef. *Journal of Food Science* 70:359–362.

Pagán, R., Condon, S., and Sala, F.J. 1997. Effects of several factors on the heat-shock-induced thermotolerance of *Listeria monocytogenes*. *Applied and Environmental Microbiology* 63:3225–3232.

Peleg, M., and Cole, M.B. 1998. Reinterpretation of microbial survival curves. *CRC Critical Reviews in Food Science and Microbiology* 38:353–380.

Reichart, O. 1994. Modelling the destruction of *Escherichia coli* on the base of reaction kinetics. *International Journal of Food Microbiology* 23:449–465.

Stasiewicz, M.J., Marks, B.P., Orta-Ramirez, A., and Smith, D.M. 2008. Modeling the effect of prior sublethal thermal history on the thermal inactivation rate of *Salmonella* in ground turkey. *Journal of Food Protection* 71:279–285.

Stephens, P.J., Cole, M.B., and Jones, M.V. 1994. Effect of heating on the thermal inactivation of *Listeria monocytogenes*. *Journal of Applied Bacteriology* 77:702–708.

Taormina, P.J., and Beuchat, L.R. 2001. Survival and heat resistance of *Listeria monocytogenes* after exposure to alkali and chlorine. *Applied and Environmental Microbiology* 67:2555–2563.

Tuntivanich, V., Velasquez, A., Orta-Ramirez, A., Ryser, E.T., Marks, B.P., and Booren, A.M. 2005. Enhanced thermal resistance of *Salmonella* and microstructure observations in marinated whole-muscle turkey. IFT Abstract 89F-34. *Presented at the Institute of Food Technologists Annual Meeting*, New Orleans, LA, July 2005.

Veeramuthu, G.J., Price, J.F., Davis, C.E., Booren, A.M., and Smith, D.M. 1998. Thermal inactivation of *Escherichia coli* O157:H7, *Salmonella senftenberg*, and enzymes with potential as time–temperature indicators in ground turkey thigh meat. *Journal of Food Protection* 61:171–175.

Velasquez, A., Tuntivanich, V., Orta-Ramirez, A., Booren, A.M., Marks, B.P., and Ryser, E.T. 2005. Enhanced thermal resistance of *Salmonella* in marinated whole-muscle vs. ground pork. IFT Abstract 89E-14. *Presented at the Institute of Food Technologists Annual Meeting*, New Orleans, LA, July 2005.

Wesche, A.M., Marks, B.P., and Ryser, E.T. 2005. Thermal resistance of heat-, cold- and starvation-stressed *Salmonella* in irradiated comminuted turkey. *Journal of Food Protection* 68:942–948.

Young, L.L., Garcia, J.M., Lillard, H.S., Lyon, C.E., and Papa, C.M. 1991. Fat content effects on yield, quality, and microbiological characteristics of chicken patties. *Journal of Food Science* 5:1527–1528, 1541.

Appendix A: Objectives and Critical Elements of Thermal Processing of Ready-to-Eat Meat Products

Erwin Waters†

Verification of Final Internal Temperature in Sausage Products

Insuring that meat products have reached the correct final internal temperature is of the utmost importance for food safety and to meet HACCP requirements.

The production and quality control (QC) departments must have the tools available to guarantee that the required internal temperatures are and have been achieved. The meat industry can no longer treat the thermal processing of meats as a secondary issue in the processing cycle and needs to pay more attention to this phase of the process.

1 That all of the products being processed meet the required temperature and time requirements.

The important part of this requirement is the word "all," which means *every* frankfurter or any other product in the batch or passing through the processing oven.

†Erwin Waters died December 2001.

Taking the internal temperature of one, or even four pieces of product, does not insure an even finishing temperature unless the equipment performs correctly and the temperatures and air distribution within the cooking chamber are even.

The process applied will have a direct effect on the product temperatures. Rapid processing schedules will result in a greater temperature variance than correctly timed processes. A steam finishing cycle, with an appropriate holding time, will result in very even finishing temperatures.

2 That the internal temperature of the product is taken at the physical center of the product, and not in any other part of the meat product.

The correct insertion of the thermometer or the temperature probe into the center of the product is of the utmost importance. The probe should always be inserted into the product at the center of the diameter of the product parallel to the length of the product, and never through the side or through the circumference of the product.

In products that do not have a formed even shape, such as primal cuts, the probe tip needs to be in the largest diameter center of the product. The most accurate temperature measurements will be obtained if the tip of the probe is equally spaced from the exterior surface of the product in all directions from the tip.

3 That the thermometer or temperature probe being used for this purpose is accurate, and that the accuracy is periodically checked and calibrated for accuracy.

The calibration of all thermometers is of the utmost importance. No thermometer will retain its calibration indefinitely and needs to be rechecked periodically. The calibration must be done often and on a predetermined schedule, with the results accurately documented.

4 That the thermal process is accurately recorded and documented.

The accurate recording of the dry bulb, wet bulb, and internal temperature of the product during the total processing period is of the utmost importance. The records are important for food safety documentation, product recall requirements, and to enable management to adjust processing cycles to achieve the best procedure to insure food safety without excessive processing losses and processing time requirements.

5 That tools are available to the QC department to determine if the product was in fact cooked to the specified temperature with the appropriate holding period.

The tools that the QC department needs are thermometers which are accurate, easily readable, and which are calibrated on a daily basis. Chart recorders that record the time–temperature process through cooking and chilling are ideal. This information can then be used to determine compliance to requirements and calculate lethality.

The use of equipment for testing phosphatase to recheck if the product, after it has been cooled or chilled, was subjected to sufficient time and temperature treatments. With correct programming, this type of testing equipment can quickly point out if a product was cooked to the desired temperature and determine if the product should be passed for shipment. These requirements are not difficult to achieve if the correct procedures are applied.

A Insuring that all of the products in a batch achieved the same desired core temperature.

There are two primary requirements that need to be considered.

1 That the energy content and the temperature of the air in the oven are equal throughout the oven chamber.

Testing the oven chambers for equal air volume, air velocity, and air temperature distribution is mandatory. The tests need to be carried out at ambient temperatures and at the elevated temperatures.

When air is heated, it expands. The cubic feet of air per pound of air increase significantly. This makes each cubic foot of air lighter. The circulation fan exerts the same energy to both the cool and the hot air. The volume of air delivered into the oven chamber by the circulation fan is the same for cool and hot air. The pounds of air the fan delivers into the oven chamber varies in accordance to the temperature of the air. Controlling the flow of hot air is much more difficult than controlling the flow of the heavier cold air.

It is for this reason that the equilibration of airflows inside the oven chamber must be conducted at both ambient (cool) and hot temperatures. The temperature and the velocity of the air at various points inside the oven chamber need to be determined. If significant variations exist, mechanical corrections must be made. The correction is made by equalizing the volume of air being delivered into the oven chamber at all points of the delivery system, on both sides of the oven.

2 That the all of the products within the batch are of the same temperature, or as equal as possible, at the start of the process.

During the time required to prepare product for placement into the oven, the temperature of the product continuously changes. This becomes especially evident during the staging process for larger volume ovens. The difference in temperatures between the first wagon loaded and the last will carry through the process, unless the load is equilibrated at the beginning of the process cycle.

Small diameter products will be more susceptible to internal temperature variances than large diameter product. Differentials in skin temperatures are also of importance, and if the skin temperature varies significantly, the final internal temperature can also vary.

To insure that these temperature differentials do not affect the final internal temperature, an equilibration cycle should be used at the beginning of each processing program. In this cycle the fresh air and exhaust air damper are closed, no wet bulb or humidity is set or added, for small diameter products the air speed is at a low setting (if possible) and the dry bulb temperature is set at between 145 and 155°F.

This cycle is run for a sufficient amount of time to warm up the structural confines of the oven and the surface temperatures of the product. The moisture on the surface of the product will evaporate into the air, raising the moisture content of the air, which starts the surface drying process. The slowly increasing wet-bulb temperature (air humidity) will automatically equilibrate the product temperatures. The time necessary for this equilibration step will not significantly elongate the total processing time, but any additional time required is well worth the results obtained.

B Insuring that the internal temperature of the product is taken at the physical center of the product.

For large diameter products, this usually poses very few problems, but unless care is taken, a reading outside the very center of the product can occur.

Small diameter products can pose a problem, if the wrong size probe is used, and because there is not sufficient product mass to hold the probe securely in the center.

1 Positioning the probe in larger diameter products.

A product which has a cross-sectional diameter of 4 inches, has its central axis 2 inches away from the surface, and stretches from one end of the product to the other. It is along this line that the core temperature must be measured to insure accuracy.

Because energy is absorbed by the product equally from any of its surfaces, the ends of the product will also absorb energy and transfer the energy into the interior of the product. The energy absorbed into the product travels at the same speed to the interior of the product from any surface and is additive. To insure that the temperature measured in the core of the product is not the additive energies absorbed from a number of surfaces, the tip of the probe must be on the centerline and at least one diameter distance from the ends of the product.

With a 4-inch diameter product, the probe tip must be a minimum of 4 inches away from ends of the central axis (i.e., ends of the product). If the probe is sufficiently long enough, it is more accurate to insert the probe from the ends of the product, because the centerline of the product can be accurately determined.

Inserting the probe at right angles to the surface of the product, or at a slant, has a number of disadvantages and requires much more care to insure accuracy.

a The probe tip must be at the axial center of the product. Accurately judging this distance by eye is very difficult.

b Probes are usually much longer than the diameter of the product. A major portion of the probe will then be exposed to the processing medium. The metallic probe will absorb the energy from the processing medium and transfer the heat to the temperature measurement nodule in the tip.

c If a large portion of the probe is outside the meat, the weight of the probe handle and connecting wire can cause the probe to lever out of position during the process, before the protein solidified to hold the probe firmly.

d If the probe is inserted into the product at an angle, to allow for a greater depth of insertion, judging the distance to the axial center is more difficult.

2 Positioning for small diameter products.

Small diameter products present a problem because at the initiation of the processing program the product's inability to hold the probe centered correctly. Because of this fact, the probe needs to be inserted, or reinserted, into the product just before the finishing cycle. At this point of the process, the meat has congealed and is able to hold the probe securely. There are two negative aspects to this requirement.

a The rise in core temperatures during the total process is not recorded. The rise of the core temperature during the process is an important factor in determining if the program parameters are correct for the product and allows for analyzing the program to determine if adjustments need to be made.

b The necessity to insert, or reinsert, the probe just before the final finishing cycle, requires stopping of the process. Any delays during this period can cause the product to cool and have excess cooking losses. With standard probes, there is no option but to insert, or reinsert, the probe just before the final finishing cycle to achieve accurate core temperature measurements. Special needle nose probes are available, which when inserted correctly can be inserted into the product at the very beginning. Just as in the large diameter products, careful insertion into the axial center of the product is of the utmost importance.

C Periodic check and calibration of probes for accuracy.

Regardless of the type of temperature probe used, the first and most important requirement is to determine if the probes are accurate. A written procedure needs to be issued that will clearly spell out the calibration procedure, how often it needs to be done and how the results of the calibration are documented.

The probes must be checked and calibrated at ice water temperature (32°F) and at a hot water temperature (150–165°F) to insure accuracy. If adjustments are necessary, the instructions from the temperature indicator's manufacture need to be accurately followed. After the adjustments are made, the test must be repeated.

Hand-held thermometers must also be checked and calibrated in the same manner. Checking any temperature measurement instrument only at one temperature does not insure accuracy, especially if the test is carried out only in ice water temperatures.

How often the calibration test is to be performed depends on the type of instrument and the historical data of the tests. If it is found that the probe or thermometer readings drift very often, the tests need to be performed at the beginning of every operational shift. If the instrument tests indicate that there are no drifts in readings, or only drifts over longer periods of time, a check of the instrument needs only to be done at the beginning of the workday, to satisfy HACCP requirements.

The use of accurate laboratory type thermometers is of the utmost importance when running the calibration tests. It is recommended that more than one laboratory thermometer is used when conducting a calibration test, and that the same laboratory thermometers be used for all of the instruments used in the plant for checking cooked product temperatures. The laboratory thermometers need to be sent out for testing periodically, and when returned, accompanied with a certification. This insures that all of the instruments and thermometers have the same calibration and that the calibration is correct.

D Accurate recording and documentation of processing and final core temperatures.

Modern processing techniques and food safety requirements make circular charts, strip charts, or computer printouts of the whole process mandatory. All processing temperatures need to be recorded, including dry bulb, wet bulb, and product core temperatures.

The recording instruments must also be calibrated in the same manner as the oven instrumentation and hand-held thermometers. The calibration will also need to be documented in the same manner.

Clear and accurate chart or printout recordings can then be used for the verification process required by HACCP, determining where a process failed and to modify or improve processes.

The recordings must be clearly marked with the batch number, the date, and the product identification. Each oven or cooking tank load, in a batch processing system, must carry a separate batch number.

In continuous thermal processing systems, the use of individual electronic recording devices, that follow the product through the oven, is highly recommended to determine the internal temperature of the product as it moves through the oven.

Although a manual check of the final core temperature of a process batch is mandatory and highly recommended, the information gathered from recordings and the instrumentation of the ovens is more accurate. Testing the temperature of products at the end of the cycle, prior to showering, is difficult at best and requires the opening of the oven doors. Opening of the oven doors immediately allows the temperature inside the oven to drop. Small diameter product will quickly cool, and incorrect temperature readings are most often the result.

Accurately documenting the manual temperature reading and comparing it to the temperatures obtained from the oven instrumentation and the recording chart, will not only satisfy the monitoring and/or verification requirements of HACCP but will also allow for a determination if modifications of the process are required.

E Tools available to the QC department for additional verification of internal temperatures.

The QC department needs to be involved on an ongoing basis in the internal temperature verification process, both in physically checking the temperatures of products after the finishing cycle and in confirming that products that are ready for shipment were in fact cooked to the minimum required temperature.

The three tools available to the QC department are as follows:

1 Accurate, easy-to-read, and daily calibrated hand thermometers. Electronic thermometers with wire leads are recommended. These can be used by inserting the probe into the product, allowing the wire to pass through the door closure point, and reading the temperatures of the product with the oven doors closed. The probe can be inserted into the product just prior to the finishing cycle, and readings taken during and at the end of the cycle. This will result in a much more accurate result than trying to take temperatures with the oven doors open.

2 The records kept by the oven operational personnel need to be examined by the QC department to determine if the temperature goals have been met and to check the results obtained by the manual temperature measurements taken by the QC department. The documentation of these investigations will form a historical database on which the accuracy of the process can be verified.

3 The equipment for measuring phosphatase measures the residual uncooked meats in a product at any time even after chilling and storage. The most, user-friendly equipment for this purpose is a system supplied by Charm Sciences and is called the "Chef Tests." The testing procedure is attached to this presentation.

Using this testing procedure, the QC department can verify that products have been correctly cooked at any point of the process after showering, chilling, packaging, storage, and just prior to shipping. Following the control point procedures recommended by Charm Sciences, very accurate verification results can be obtained.

Critical Controls for HACCP and Listeria Requirements

Temperature Tracking Equipment

Bacteria are natural in the environment and raw meats are no exception. Most of the bacterial contamination is on the surfaces of raw meats. Comminution and injection will take the surface bacteria and mix it or spread it into the whole product. Inadvertently, additional bacteria could be added to the product during the preparation stage of the meat for any ready-to-eat products, whether comminuted or injected. We must take it for granted that pathogens are present. These must be deactivated with a correctly carried out cooking process.

The FSIS has scientifically determined, and is insisting on, that for food safety a 7-log kill is achieved during the cooking process. This requirement is the first, and primary, critical control point for ready-to-eat meat products. The FSIS also now requires that all processors determine if the 7-log lethality is achieved during the cooking process for each of the products manufactured. Even though the FSIS has issued a "Safe Harbor" method for the cooking of meat products, each processor should, and must, check the cooking method applied to insure that in fact the minimum allowable lethality is achieved.

Running process lethality determinations is really not difficult especially if the "Safe Harbor" cooking methods are incorporated into the process. If the cooking process is completely out of range from those suggested in the "Safe Harbor" methods, the lethality determination process becomes more difficult, because challenge studies might have to be performed. For standard ready-to-eat meat products, there should be no reason for not adhering to the "Safe Harbor" methods. Some people will at the present time insist that it is not necessary to conduct process lethality determinations if the "Safe Harbor" methods are applied. Knowing how government regulations evolve, this statement could be false.

The process lethality determination should be done in two phases. The first phase will determine if the cooking program being applied conforms to lethality requirements. To carry out this test, three tools are required.

1. A computer, almost any PC will do.
2. The Process Lethality Determination computer program, which is available free from the AMI web site.

3 An accurate time/temperature recording instrument that simultaneously records from start to finish the internal core temperature of the product and the time. Atkins Technical manufactures a very good and inexpensive instrument for this purpose, including software for automatic downloading into the computer.

Using the computer software, the instrument is programmed to record the temperature at any desired time interval. The probe is placed into the core center of the visually largest product. The product is moved into the cooking oven, the small hand-held microprocessor is attached to the probe with a sufficiently long wire to be kept outside of the oven. The instrument is activated and records the process. During the process the operator can check the core temperature at any time from the instrument.

At the end of the cooking process, the instrument can be left with the product to record the chilling time–temperature relationship or it can be removed for information processing. The instrument is connected to the computer, and the recording is downloaded. A hard copy of the times and temperatures, as well as a graph is obtained. The lapse time and temperatures are then transferred to the process lethality program and a theoretical, but accurate, log kill verification can immediately be determined.

If the first phase indicates that lethality has been achieved, the second phase of the verification process is to repeat this test but include two laboratory tests.

1 Send a sample of the product being tested to a recognized laboratory for microbial testing.
2 Immediately after cook completion, and prior to removal from the oven, take a sample of the product for microbial testing. If done correctly, this eliminates any possible cross-contamination after cooking.

The information obtained is the first step for the restructuring of the HACCP plan for Listeria control. This information will also give the processor a complete new view of the cooking process and the adjustments that could be possible for improvement of food safety and shelf life.

Cross-Contamination Possibilities

Even though the FSIS is focused on *Listeria monocytogenes*, any bacterial cross-contamination of the product prior to entering the postcooking areas needs to be addressed. If the cooking process was carried out correctly and sufficient lethality was achieved in the cooking process, any bacterial contamination of the clean areas must come from other sources than the interior of the product. Recognizing from where cross-contamination can be transferred to the clean areas is the primary step in controlling the possibility of cross-contamination.

1 The room in which the ovens are located, and the methods applied during the oven unloading process, can present the greatest danger to cross-contamination of the surface of the product, after it has been pasteurized during the cooking process.
2 The sanitation of the persons working in the clean areas is the next most probable reason for the cross-contamination of the structures and the equipment used for further processing of the clean areas.
3 Clean area environment and bacterial transfer locations.

The transfer of bacterial contamination into the clean areas could be caused by any one, or a combination, of these factors. Conducting swab tests, after sanitation, during pre-op inspection will only indicate if the sanitation crew performed their job correctly but will not indicate if cross-contamination is taking place during the workday because of any of these three factors. If the pre-op tests were negative, taking swab tests at the end of the shift or workday, prior to sanitation, will immediately indicate if cross-contamination has taken place. Finding the specific cause of cross-contamination might be like finding a needle in a haystack, but it is the only logical method that can avoid repetition of this problem.

Testing the "clean area environment and bacterial transfer locations" (item number 3 above) is a relatively simple but time-consuming process. It needs to be conducted when the clean area operation is closed for more than one day (e.g., over a weekend). A thorough sanitation of the areas should be carried out. When sanitation is completed, being very careful not to recontaminate the rooms, and swab tests are taken of the most commonly contaminated areas, such as refrigeration drip pans, floor drains, heavily used contact area, rail supports, ceiling, and wall areas over equipment or work stations.

The refrigeration equipment and any positive air pressure equipment should be left running to maintain the rooms at the same condition as during normal operations. Prior to any person's entry into the area, and prior to pre-op inspection, one person from QC will repeat the swab tests on the same areas taken prior to weekend shut down. If these tests are negative, it can be taken for granted that environmental cross-contamination is not an immediate problem. It is recommended that this test be performed three times in a row and then quarterly.

If any of the morning tests are positive, the area where a positive reading is indicated will need to be closely examined to determine the source and cause of contamination. The investigation will need to focus on the most common probabilities first, such as the products if the positive results came from an area where products are stored, the air and air circulation equipment in the room, structural parts of the equipment in the room, and the floor drains.

Floor drains are of major concern because the drainage system connects to other drains in the plant area. The floor drain concentrates all of the water being used in the room and in the investigative period. Testing of drains during the workday is a good system to determine if cross-contamination is taking place. If bacterial contamination is not being introduced into the clean area by personnel, product, ambient air or transfer through the sewer lines, then the drains should remain bacterially clean from the morning to the end of the workday.

Swab tests of contact surfaces prior to sanitation and microbial tests of the surfaces of products will form a database from which the causes of the cross-contamination of the clean areas can be better determined.

Additional Information on Cooking and Chilling

Appendices A and B (USDA FSIS, 1999a, 1999b) were issued by the FSIS and they detail the cooking and chilling requirements to attain the prescribed lethality. It is important to note that cooking in itself will not satisfy the regulations, chilling of the products must also conform to requirements.

Chilling of large diameter products takes longer than cooking. The larger the diameter of the product the more difficult it is to achieve.

Management Education

Management must understand all of the mechanical and process parameters related to the above necessities. The educational process must include the following information.

1 Preparation of products for thermal processing:

a Even loading of products on the processing carts.

b Determining the size differential of products that can be processed together.

- **c** Correct insertion of control and tracking probes into products to achieve a continuous recording of the process.
- **d** Surface condition and cleanliness of products for slicing prior to heat processing. This eliminates excess protein contamination of the product surfaces which can then become a bacterial breeding ground.

2 Thermal requirements and factors for cooking:

- **a** Processing qualities of air: dry and wet-bulb temperature relationships, air velocities, energy transfer science, air volume as compared to air temperature, and evaporative cooling.
- **b** Processing quality of water in water cooking operations: water temperatures, availability of energy from water, stratification of water temperatures, and the necessity of keeping a good agitation in a water cooking system.
- **c** Heat transfer from cooking medium to product: the effect of heat transfer by the cooking medium's energy and velocity.
- **d** Heat transfer from the surface of the product to the product's interior.

3 Cooking equipment requirements:

- **a** Energy requirements.
- **b** Air or water velocity requirements.
- **c** Capability of wet-bulb control.
- **d** Air or water distribution.
- **e** Accurate controls for both time and core temperature-based processes.
- **f** Accurate and easily readable recording equipment.
- **g** Construction of equipment for easy sanitation and elimination of possible cross-contamination: floor drains in ovens, fresh air intakes for both hot and cold processing, and ability to adjust for accurate air distribution.
- **h** Maintenance requirements to assure consistent operation.
- **i** Installation requirements to avoid bacterial breeding locations.

4 Process requirements to achieve lethality:

- **a** Achieving core temperature increases through the danger zones in the minimum amount of time.
- **b** The relationship of the wet-bulb temperature (humidity) in achieving this goal.
- **c** The effect of wet-bulb temperature on lethality.
- **d** Holding time at the end of the cycle to insure minimum lethality and the possibility of increasing lethality.
- **e** Rethinking of low temperature, stepped, processing programs which extend the time that the product is subjected to temperature danger zones.

f Equilibration of the product to insure that all of the products attain the minimum planned internal temperature at the end of the process.
g The use of liquid, or sawdust-generated smoke and their effects on the surface cleanliness of the product.
h The importance of the use of a cooking, or high humidity, cycle in the final step of the process.

5 Showering:

a Shower systems to avoid surface cross-contamination after cooking.
b Water quality determination.
c Duration of showering, or to what internal temperatures to shower to, in order to achieve best chilling time.
d Sanitary requirements for external and in-oven showering systems.

6 Chilling of ready-to-eat meats:

a Mechanical requirements for chilling of both small and large diameter products: energy and air velocity requirements, construction, and controls.
b The use of brine chillers for small or large diameter products. Brine quality and cleanliness.
c Avoidance of cross-contamination during the chilling period from exterior sources and from floor drains and equipment inside the room.
d Thermodynamics of chilling. Heat transfer and removal.
e Tracking of chilling process to determine required process changes or increases in refrigeration energy or air volume.

7 Avoidance of product surface cross-contamination after cooking:

a Oven room operations: division between loading and unloading of ovens and water-cooking equipment.
b Personal sanitation of employees in oven rooms and chillers.
c Sanitation requirements prior to unloading an oven or water tank.
d Control of traffic in oven room during unloading.
e Floor drain sanitation in oven room and chillers.
f Cross-contamination of chilling rooms through loading procedures.
g Sanitation of chilling rooms.

8 Postpackaging pasteurization:

a Heating of packaged product to achieve product surface lethality of any possible cross-contamination.
b Systems available.
c Process for individually packaged products or multiple packaged products.
d Temperature and time requirements.
e Solid muscle product systems.

f Immediate chilling requirements after postpackaging treatment.
g Shelf-life expectancies and measurements.
h Temperature verification processes.

9 Changes in processing to minimize cross-contamination:
a Solid muscle items bag cooking. No removal from bag.
b Hot packaging of solid muscle items and surface sterilization in shrink tunnel.
c Sterilization of plastic or impervious casings prior to stripping for slicing.
d Temperature tracking from start through chilling.
e Calculating lethality.
f Laboratory tests to confirm lethality.
g Arrangement of records and record samples.
h Inclusion of system into HACCP documentation.

References

USDA FSIS. 1999a. *Appendix A: Compliance Guidelines for Meeting Lethality Performance Standards for Certain Meat and Poultry Products.* Accessed at: www.fsis.usda.gov/oa/fr/95033f-a.htm.

USDA FSIS. 1999b. *Appendix B: FSIS Compliance Guidelines for Cooling Heat-Treated Meat and Poultry Products (Stabilization).* Accessed at: www.fsis.usda.gov/oa/fr/95033F-b.htm.

Index